Iowa's Remarkable Soils

T0289530

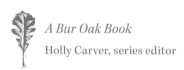

A Bur Oak Book

Holly Carver, series editor

IOWA'S REMARKABLE SOILS

The Story of Our Most Vital Resource and How We Can Save It

Kathleen Woida

UNIVERSITY OF IOWA PRESS | IOWA CITY

University of Iowa Press, Iowa City 52242

Copyright © 2021 by the University of Iowa Press

www.uipress.uiowa.edu

Printed in the United States of America

Design by April Leidig

Printed on acid-free paper

Library of Congress Cataloging-in-Publication Data

Names: Woida, Kathleen, author.

Title: The Story of Our Most Vital Resource and How We Can Save It / Kathleen Woida.

Other titles: Bur Oak Book.

Description: Iowa City: University of Iowa Press, [2021] | Series: Bur Oak Books | Includes bibliographical references and index.

Identifiers: LCCN 2020030541 (print) | LCCN 2020030542 (ebook) | ISBN 9781609387501 (paperback) | ISBN 9781609387518 (ebook)

Subjects: LCSH: Soils—Iowa. | Soil ecology—Iowa. | Soil Conservation—Iowa.

Classification: LCC S599.18 W65 2020 (print) | LCC S599.18 (ebook) | DDC 631.409777—dc23

LC record available at https://lccn.loc.gov/2020030541

LC ebook record available at https://lccn.loc.gov/2020030542

In appreciation of the dedicated soil scientists, cartographers, and technicians who labored over many decades, mostly in obscurity, to study, classify, and map the soils of Iowa. The immense natural resources database they created and continue to update is unparalleled in its usefulness to the citizens of this state.

Contents

ix   Acknowledgments

xiii   Preface

1   Introduction

PART ONE

The Inheritance: Fertile Black Gold

ONE

11   Profiles of the Underground

TWO

37   Wealth in Diversity

THREE

61   The Stories They Can Tell

FOUR

85   Soils on Iowa's Hidden Landscapes

PART TWO

The Sixth Factor: People, Agriculture, and Soils

FIVE

105   Reaping the Bounty

SIX

119   Squandering the Inheritance

SEVEN

139   Rediscovering the Living Soil

EIGHT

161   Stories from the Field

NINE

187   Soils, Climate Change, and the Future

199   Glossary

211   Bibliography

225   Index

Acknowledgments

The idea for this book incubated in my mind for at least three years before I began to write it, often while pondering the Iowa landscape on my drives to and from field sites all over the state. But despite my diverse experience with soils, as a geologist I knew I would need input from specialists in the many soil disciplines I hoped to address. It is only because of the assistance of many knowledgeable people from all over the state that this project ever came to fruition.

When I moved back to Iowa in 2000 after seven years with the U.S. Department of Agriculture's Natural Resources Conservation Service in the Southwest, I was privileged to work with many excellent soil scientists across the state. The most inspiring of them was Rick Bednarek, the agency's head soil scientist for Iowa. Rick and I formed a bond born of misery from fieldwork during nasty winter weather over the years. As soil health promoter for the agency in Iowa, he never failed to impress me with his enthusiasm, humor, and ability to connect with the public as we traveled the state to workshops. Jason Steele, Neil Sass, and Patrick Chase put me in touch with farmers, helped me obtain soil samples, and answered all my many questions. They meet regularly with farmers in their fields to spread the word about soil health and regenerative agriculture where it has the most impact. And nothing illustrates the dedication of Iowa's soil scientists better than the generous donation the Professional Soil Classifiers of Iowa made to the nonprofit University of Iowa Press to offset publication costs of this book in the hope of making more readers aware of the important issues surrounding Iowa's soils.

As the writing progressed, Jerry Miller, professor emeritus of soil science at Iowa State University, wholeheartedly offered to review as many chapters as I wanted to send him. That ended up being nearly all of them. Jerry's words of encouragement motivated me to keep going, and his corrections and suggestions have made this book so much better than it would have been otherwise. Michael Thompson, professor of soil science at Iowa State with an expertise in soil genesis, soil chemistry, and paleosols that is second to none, closely reviewed all of part 1 and clarified many points. In the 1990s, Michael had graciously agreed to serve on my dissertation committee and welcomed me to use his lab and specialized microscope in the Department of Agronomy, all the while sharing his vast knowledge of soils. To him more than anyone, this geologist owes her improbable discovery of the science of pedology and the prominent place of soils in the evolution of landscapes.

For the latest research into Iowa's glacial geology, I turned to Stephanie Tassier-Surine, lead Quaternary geologist at the Iowa Geological Survey, who kindly reviewed most of part 1. Jean Prior Sandrock made a number of valuable suggestions from her broad knowledge and her extensive experience writing about Iowa's geology for the general reader. Phil Kerr of the Iowa Geological Survey patiently helped out with radiocarbon dates, and any flaws in that regard should be attributed to me rather than Phil. And Art Bettis and Dick Baker, both University of Iowa professors emeriti, clarified some issues for me regarding Quaternary paleoenvironments.

I received many helpful suggestions for chapter 5 from Iowa State history professor Jeff Bremer, who is writing an updated history of Iowa. For the chapter on soil erosion and degradation, I turned to Iowa State professor Rick Cruse, an expert on soil management and leader of the Daily Erosion Project, whose input was instrumental to what is a major theme of this book. Tom Loynachan, Iowa State professor emeritus, graciously reviewed the essential material on soil biology and microbiology. Doug Peterson, Natural Resources Conservation Service soil health specialist for the central United States with practical knowledge and indispensable field experience with farmers, reviewed chapter 8 on soil health and regenerative agriculture. Marshall McDaniel of Iowa State's Department of Agronomy, Jerry Hatfield of the National Laboratory for Agriculture and the Environment, and Nick Ohde of Practical Farmers of Iowa reviewed the final chapter from their respective viewpoints.

Others who made important contributions to this book include Iowa State's Bradley Miller and Meyer Bohn, both of whom reviewed the material on soil classification. Meyer produced much-needed updated maps of Iowa's six soil orders while steadfastly tolerating all my annoying suggestions. Rebecca Kemble, friend and artist, created in oils my evolving image of a composite soil profile to capture the soil horizons found in Iowa. I am most indebted to all those who contributed photos to this book, which I believe will increase its usefulness many times over. Many thanks to Shirley Ten Natel and Angela Dekkers in Ireton, Iowa, who went to much trouble to provide me with a digital version of the rare 1890s photo of farmers breaking prairie in northwest Iowa. I also wish to thank my former colleagues in public affairs at the Iowa Natural Resources Conservation Service, Jason Johnson and Laura Crowell, who patiently answered my emails as I dug for various facts and figures. Ryan Dermody of the Waverly Soil Survey office provided extensive data on the cadre of Iowa soil mappers since 1901, and Jennifer Welch of the Polk Soil and Water Conservation District shared some much-appreciated artwork.

Chapter 8, which raises hope for the future, was possible only because of the kindness of those Iowans who took time out of their busy lives to meet with me and share their accounts of new ways of farming and caring for the land. The passion and commitment of these modern pioneers and of the soil scientists I interviewed strengthened my resolve to bring their stories to light.

To Catherine Larson, friend, librarian, and professional indexer who generously offered to create the book's index under a tight deadline, thank you very much for saving me the trouble and debacle of attempting to do it myself. Sincere thanks go to my longtime friend Merle Shapera, a true city-loving Chicagoan who may now know more about soils than she ever wanted to. And I am more appreciative of the sustaining strength of family than any of them realize, which goes well beyond this project in time and scope. To Ben, for being a wonderful model of strength, determination, and excellence in all that you do, thank you, son. I am grateful for my older sister's interest in the book and, in fact, may have been fated to learn and write about soils and soil health in particular because of her. As a child I had always been fascinated by a story about Mickey as a toddler, sitting in the field in a potato crate, eating the fertile loam with a spoon. The fact that she is now in her early seventies, healthy as a horse, and an avid gardener must tell us something! In a more serious vein, though, I believe having roots in a family that practiced old-fashioned agriculture for generations—strong on human labor and appreciation of the land—may have laid the groundwork that ultimately led to this book.

I owe my lasting gratitude to Holly Carver, Bur Oak Books series editor at the University of Iowa Press, for having faith in me in the first place and then expertly guiding me along the path to completion, always with a welcome dose of humor. My heartfelt thanks also go to Karen Copp, Meredith Stabel, Jacob Roosa, and Susan Hill Newton at the University of Iowa Press, who worked their magic to transform my vision into vivid reality. I can't imagine having a better team to work with.

Preface

Iowa may have the best soils on Earth, and yet so few of its citizens really know much about them. You can find a number of excellent books about Iowa's prairies and woodlands, our birds, our wildlife, our landscapes, and other natural resources. But very little is available outside of the scientific literature about Iowa's soils, despite their undeniably critical role in the economy and environmental quality of our state. This verdant and productive land between two rivers has been my home for nearly thirty years. During that time, I've had Iowa soil under my fingernails most days, at work or at home, and when I didn't I was probably thinking or writing about the land for one reason or another. In the pages of this book, I hope to share with you what I have come to know and cherish about the soils I've had the good fortune to contemplate.

Although not a native Iowan, I come from a farming background. My great-grandparents homesteaded our family farm in Michigan in 1873, and for generations it has supported potatoes, grain, hay, an orchard, a large vegetable garden, tree plantings, and livestock. Much of the farming took place before the days of high-tech machinery, when the work entailed walking the fields in close contact with the soils, the roots, and the earthworms. One of my clearest memories from childhood is heading off to the fields after school with five or six siblings to walk beside a wagon for a couple of hours, "picking stones." It was an inescapable spring chore on farms in the northeastern Lower Peninsula of Michigan, where thin glacial deposits cover Devonian limestone bedrock, and where winter frost inevitably heaves rocks to the surface. Our fields were dotted with tall stone piles where we dumped our rocky harvest. They were a testament to the maddening, relentless churning of the soil each and every winter since my ancestors laboriously transformed the farm from woods to fields 150 years ago.

Life on the farm ingrained in me a deep love for landscapes and how they came to be, an appreciation that only grew deeper with the passing years. After college and an earlier career, I completed a Ph.D. in geology at the University of Iowa, studying a remarkably thick buried soil—a paleosol—that lay at the land surface in southern Iowa for a few hundred thousand years before being buried by silty loess. I then went to work for the U.S. Department of Agriculture's Natural Resources Conservation Service, first in Utah and New Mexico and finally in Iowa for the last twenty years of my career. Called the Soil Conservation Service until 1994, it was born in the 1930s following several years of disastrous droughts and tremendous soil erosion on the Great Plains. In Iowa, I worked closely with the agency's soil scientists and engineers and came to appreciate the exceptional soils that blanket this state, particularly when compared to the thin and infertile soils I had seen in the Southwest. And unlike our Century Farm in northern Michigan, very few Iowa soils require annual harvesting of stones. Most have wonderful silty textures, perfectly suited to planting and growing crops and capturing and holding rainwater to sustain plants. During my years with the Natural Resources Conservation Service, I came to believe that Iowa's soils truly represent this state's most precious and fundamental natural resource.

Unfortunately, I also saw firsthand the damage that has been done and continues to be done to this superb resource. In writing this book, I hope that I can bring to light what an irreplaceable treasure the soils of Iowa are by explaining in a layperson's language the unhurried and complex way they came to be, the stories they can tell us about Iowa's prehistoric past, the exceptional condition they were in before widespread cultivation, the degraded condition of many of them today, and the ways in which more and more forward-looking farmers and property owners are halting that degradation and actively improving the biological health and physical quality of their soils.

My desire is that this book will be useful and thought-provoking to a wide range of people. With that in mind, I have included just enough technical detail to do justice to a complex resource but not overwhelm the average reader; I have also included a glossary of terms. So whether you are someone who works with the soil, a student of the environment, or someone who knows little about the topic but cares about Iowa's future, I hope that this story of our soils gives you a deeper awareness of their nature and vital importance.

A few notes for clarification. Labels on the photographs of soil profiles are my interpretations and not those of the photographers. For most measurements, this book uses the English system of inches, feet, acres, and so on in order to be more accessible to the general American audience. However, scientists in the United States and around the world uniformly use the metric system in their work and publications, and I have used the metric scale to express the size of tiny things like sand, silt, and clay particles and minuscule features in photographs taken with a microscope. I believe it is simply easier to visualize a dimension of half a millimeter, for example, than two one-hundredths of an inch. Finally, some of the radiocarbon dates presented in this book are older than those originally reported in the scientific literature. Beginning in the 1990s, scientists began calibrating existing conventional radiocarbon dates for greater accuracy.

Introduction

IT HASN'T ALWAYS been the case, but very few of us today pay much attention to the ground beneath our feet. Most urban dwellers—and even some rural residents—spend most of the time indoors. When we do venture outside, we're usually walking on concrete or pavement. Even if we realize that most of our food comes from the soil, it's just not something of great interest, and most of us simply take it for granted. Certainly, there are exceptions. Farmers and gardeners appreciate the importance of soils to their crops and gardens, engineers must consider what's beneath the surface before they design and build something, and quarry operators are acutely aware of what it costs to remove the overlying sediments to get to the economically worthwhile limestone. There are other examples. But for most people, the soil has always been there and always will be. Why worry? If the developer stripped it off in my subdivision, can't I just buy some topsoil and sod and replace what nature left there? And my crops seem to be growing and producing pretty darn well despite erosion, as long as I apply enough ammonia.

In fact, soils are of critical importance to our society and to civilization, and rather than being deposited by nature, it took thousands of years for them to develop, right there in the same spot where you find them today. Other books have appeared in the last few years addressing the peril of soil loss and degradation around the globe, but this isn't a new discovery. In 1953, W. C. Lowdermilk submitted a compelling thirty-page report to the U.S. Department of Agriculture called "Conquest of the Land through 7,000 Years," in which he illustrated how soil erosion and land degradation had toppled empires and wiped out civilizations. It was a warning driven by the devastation of the Dust Bowl on the High Plains in the 1930s. In this book I shine a light on Iowa, one of the most productive agricultural areas of the world that started with the richest soils on the planet. But Iowa also has the most transformed landscape of any state in the country, and that conversion has altered its soils in far-reaching ways.

Historically, agriculture has been the lifeblood of this state. Iowans have been called country bumpkins by the rest of the country for so long that many now claim the snub with pride, boasting about the ten-foot-high corn and the "largest in the land" state fair. In some ways, however, the state has become less rural in recent decades. If we use population as a yardstick, nearly two-thirds of Iowans now live in urban areas rather than on farms or in small towns (Iowa Community Indicators Program 2020). Yet despite this demographic shift, agriculture still plays a huge role in Iowa's economy, with agriculture and related businesses accounting for more than a quarter of the state's gross domestic product as of 2017, the most of any one sector. Even more importantly, if we use acreage as a measuring stick, agriculture is still the state's indisputable king. In 2017, 81 percent of Iowa's land was dedicated to crops, grassland cut for hay, and pasture, more than in any other state in the country. At the same time that most of us are proud of our state's agricultural prominence and feel good about "feeding the world," we worry increasingly about the effects of agriculture on our natural ecosystems, air and water quality, and health.

As it turns out, the most important thing we can do to lessen our worries and provide a cleaner environment for our children and grandchildren is to pay attention to our soils. Because what is good for our soils is great for our crops *and* our water and air. Fortunately, in this twenty-first century, it appears that slow but encouraging changes in how we treat our soils are in the works. These changes cannot come too soon. Many believe that at current rates of soil loss, our topsoil will be gone in less than a hundred years, with or without the additional effects of climate change.

To truly appreciate the importance and complexity of our soils and why we need to change our ways in order to protect them, a certain amount of knowledge about the subject at hand is essential. Many written sources about soils or agriculture for the lay reader short-change discussion of the natural history of soils, about which most people know very little. To that end, in part 1 of this book, I will focus on the major environmental and chronological factors in the formation of our soils, the amazing diversity of soils across the state, and what they can tell us about Iowa's past and its future potential—if we will only listen.

The world's most fertile soils, called Mollisols, developed under grassy vegetation: the prairies of North America, the steppes of Russia and Ukraine, and the pampas of Argentina. Not coincidentally, all these grasslands are now largely cropland and the major grain-producing regions of the world. In nearly all of Iowa and parts of other midwestern states, it was tallgrass prairie that flourished, with bluestem grasses tall enough to hide a rider on horseback. With its extraordinarily deep roots and diverse life-forms, both macroscopic and microscopic, the tallgrass prairie contributed abundant carbon to the soils year after year. For manufacturing fertile soils, it was the best of the best of the world's grasslands.

The signature characteristic of a Mollisol is its thick dark A horizon. Commonly called the topsoil, it is rich in organic matter and the nutrients that plants need, and it has a remarkable ability to store water. In the latter half of the nineteenth century, following the invention of the moldboard plow, Midwest settlers came to realize that the prairie land they thought was unsuitable for agriculture because of its dense root mass was in reality a wealth of fertility. The topsoil of the prairie was thick and dark and came to be known as the black gold of Iowa (fig. 1).

In addition to dense, deep-rooted prairie vegetation, most of the land now called Iowa was the beneficiary of another gift, one that fell from the sky. Between about 29,000 and 15,000 years ago during North America's last great ice age, sediment consisting of silt-sized mineral particles with a dash of clay and the texture of flour fell onto the landscape at a rate of a few tenths of an inch per year. When it was all done, anywhere from several inches to more than 200 feet of this amazing dust, called loess, covered the land between the Missouri and Mississippi Rivers. The word "loess" comes from a German word meaning

1. Iowa's black gold. Photograph by Lynn Betts, USDA–Natural Resources Conservation Service.

"loose," and it is exactly this friable, noncompacted quality that gives loess its superior workability and ease of root penetration, called tilth, and its ability to soak up rainwater in just the right amount to sustain prolific plant growth.

In the simplest of terms, soils consist of air, water, and two solid ingredients: mineral matter and organic matter. In most of Iowa, the combination of pliable loess (the mineral component) and the bountiful life of the tallgrass prairie (the organic component) resulted in superb Mollisols. There are several kinds of Mollisols. Those covering much of the Great Plains—parts of the Dakotas, Nebraska, Kansas, Colorado, Oklahoma, and Texas—formed under shortgrass prairie in a notably drier climate and are less rich. But the Mollisols of Iowa, which once made up more than two-thirds of its soils, more than in any other state, are the most fertile type. What's more, another quarter of Iowa's soils formed under deciduous trees, and this soil type is the world's second most productive for agriculture. Blessed with such fertile soils and an agreeable climate for growing crops, Iowa is a very close second only to Illinois in leading the country in acres of prime farmland for crops, despite Iowa's much hillier topography.

Part 2 of the book focuses on the human imprint—what I call the sixth factor of soil formation in addition to parent material, climate, organisms, topography, and time—and the changes to our state's premier soils that resulted from settlement and agriculture. When the Iowa prairie was first plowed, the settlers found 14 to 16 inches of topsoil. But by 2000, the average thickness was only 6 to 8 inches, and the remaining topsoil had lost 30 to 50 percent of its organic matter. In 1982, the average rate of erosion on cultivated land in Iowa was so high that the erosion map of the United States showed a jarring purple bull's-eye centered on Iowa and northern Missouri. By 2017, the rate had improved considerably, but Iowa still stood out as a vivid orange on the map—it had lost 150 million tons of soil that year. It also meant that Iowa was responsible for at least 15 percent of the country's annual soil loss from cropland, despite owning only 7 percent of that cropland. What's more, alarming studies in the past decade by the Environmental Working Group suggest that the soil loss represented by these numbers, which do not include gully erosion, may constitute only about *half* the total amount of soil leaving Iowa's fields.

This loss of Iowa topsoil has resulted in what we could call the browning of the black gold. As most people realize intuitively, the humus that makes soils black is also what makes it fertile. As humus is lost to erosion, there is an accompanying loss of fertility. Farmers have responded by increasing the application of chemical fertilizers that began in the 1950s. This gained steam in the 1970s when Secretary of Agriculture Earl Butz told farmers to "get big or get out" and "farm fencerow to fencerow." Changing that model has become increasingly difficult under the current system. Many farmers have invested millions of dollars in equipment to raise corn and soybeans and nothing else, corporations have bought up family farms, and more than half of Iowa's cropland is planted and harvested by tenant farmers who have little emotional investment in the fields they farm but who cannot afford to purchase their own land due to skyrocketing prices. Our current political environment pushes for more and more production, seemingly at any cost, and favors the chemical industry to increase both crop yields and the number of acres cropped.

As a result, the majority of Iowa's farmers have accepted chemical-heavy farming as the best and only way to farm. Across Iowa, synthetic fertilizers and pesticides kill off essential microbial life and mask the true depleted nature of our soils. At the same time, conventional tillage practices and improper "rotations" consisting of corn with an occasional crop of soybeans have left the soils compacted and incapable of soaking up and storing water, so that most rainfall leaves the fields as runoff instead of penetrating to the root zone of plants. Meanwhile, several studies have suggested a strong link between the presence of agricultural chemicals in our water and air and human health problems.

What's more, the erosion of our topsoil affects not only those of us living in Iowa but also populations and ecosystems downstream from our state. Nitrate from runoff and phosphorus attached to the sediment produced by soil erosion are the main culprits contributing to the expansion of the Dead Zone in the Gulf of Mexico that is greatly damaging the ecosystem and the shrimp industry.

The Gulf of Mexico Dead Zone is one example of why understanding soils is important in the applied natural sciences—environmental science, hydrology, geology, botany, wildlife biology, forestry, and so on. It's also critical to anyone working in the fields of engineering, water resources, food science and nutrition, waste management, archaeology, and several others. And, of course, knowledge of the properties and influence of soils is indispensable for anyone involved with agriculture—farming, grazing, and animal husbandry—and horticulture—the production of fruits, vegetables, flowers, and other ornamental plants.

You may or may not live in Iowa, and you may or may not work the land, but regardless of your role in the society and culture of your state, soils do matter to you. After all, with the exception of what we obtain from lakes or oceans, more than 95 percent of the food we eat either grows directly in soils or comes from animals who lived off the resources of the land, rooted in its soils. Nearly all our drinking water passes through and is purified by soils at some point in its long journey from the clouds to the river or the aquifer.

In truth, we live on the soil, we eat and drink from it, we build on and with it, and when our lives are over we are buried in it or scattered over it. This priceless and vulnerable skin of the earth is as fundamental and vital to human life as anything on our planet (Bogard 2017). The distinguished soil science professor and Quaker from Wisconsin, Francis Hole, used to say that all living things, including us, are only "temporarily not soil." Hole also maintained that "the earth beneath our feet has a mystique, which explains why people kneel and kiss the ground and call it Mother Earth." I've talked with Iowans who grew up on family farms and raised their offspring on the same land. In so many words, they've told me that for them the land is the repository that holds their family memories. In a literal way, our soils also hold in their memories physical evidence of what Iowa's landscapes have experienced in the past several thousand years.

I hope the pages that follow will convince you of the beauty and mystique of our soils. You will see the intricate architecture of a soil that is visible under a microscope—which I hope might inspire you to go outside and dig some up, to handle it and smell its distinctive earthy aroma emanating from the life within it. You will come to know the fascinating path by which our fragile soils came to be and the many ways we humans unintentionally damage them. You will also learn of the encouraging and proliferating changes beginning to take place in Iowa and elsewhere. As more and more farmers and property owners join the growing soil health revolution with innovative and sustainable practices, they just might save Iowa's greatest treasure.

PART ONE

The Inheritance

Fertile Black Gold

CHAPTER ONE

Profiles of the Underground

How can I spend my life stepping on this stuff and not wonder at it?
—WILLIAM BRYANT LOGAN, 1995

IN THOUSANDS OF FIELDS across the landscape in some twenty-eight Iowa counties, there reside patches of land several acres in size that are carpeted by a superb soil called Tama. There are nearly a million acres of this soil in Iowa, and because it is one of the state's most productive agricultural soils, it has been designated the state soil of Iowa. Tama soils can yield more than 250 bushels of corn per acre under proper management, so nearly all of them are under cultivation. They are representative of the state's many soils that formed in loess under tallgrass prairie, and most of the characteristics of these soils came into being long before they were farmed. Many of those characteristics are still visible in the Tama soil profile and in many similar soils (fig. 2).

Similar to portrait photography, where a profile means the side view of someone's face, a soil profile is the side view of a vertical section of soil where it is exposed by an opening in the earth. It could be on the side of a gravel pit or clay pit, on a steep wall of overburden—sediments overlying bedrock in a limestone quarry—or on an eroded streambank or side of a gully. Since such ideal exposures are very limited across the landscape, scientists who study and classify soils more often work from soil cores they obtain with manual or machine-operated augers and probes. Pits excavated for construction or archaeological digs also

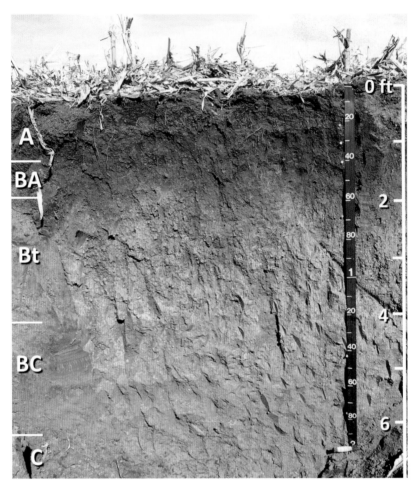

2. Tama soil profile, the Iowa state soil. Courtesy of the USDA–Natural Resources Conservation Service.

afford excellent opportunities, as well as pits dug expressly for the purpose of soil studies. For example, students studying soil science at Iowa State University and scores of high school students in Iowa's many Future Farmers of America chapters participate in soil-judging field days (fig. 3). There they climb into soil pits 5 to 6 feet deep—dug with a backhoe in pastures and fields, sometimes stabilized with wooden

3. High school students at an Appanoose County soil pit, training for an upcoming district soil-judging contest. Photograph by the author.

supports, and backfilled at the end of the day—to describe Iowa soils and evaluate their suitability for different uses.

The soil profile also has a more general meaning when applied to a group of similar soils. Each of Iowa's major soils, called soil series, has a characteristic sequence and thickness of soil layers with depth: in other words, a two-dimensional characteristic profile. Every county in Iowa has a collection of documents, called a soil survey, containing a detailed description of a representative profile for every one of the soil series in the county.

Some series occur at many locations across the state, and a few such as the Colo series can be found in most of Iowa's ninety-nine counties. Soils are closely linked to geographic place. Unlike the naming of biological or botanical species, which often include the discoverer's name in the Latin or Greek designation, soil series are typically named after towns or geographic landmarks close to the type locations where they were first defined. Some series names come from adjacent states, and some are invented. As Francis Hole once pointed out, the only way you could have a soil named after you would be to establish a village and get it named after you, then persuade a soil scientist to name a soil after your village. (As far as I know, no one has ever taken him up on that, and most soil scientists I've known are pretty unassuming.) The type location of the Tama soil series is in Tama County, about three miles north and one mile west of the town of Gladbrook. If you were to google its official series description, you would find the exact location within a few feet.

To the uninitiated, at first glance, the Tama profile in figure 2 seems to consist simply of a dark gray zone about 1.5 feet thick overlying a lighter brownish zone more than 4 feet thick. But if you could handle and examine it, with a little guidance you would see that there is a lot more going on. This profile has at least five different horizontal zones, called horizons, which are distinguished by their physical, chemical, and biological properties. Note the labels on the left side of the photograph.

However, before getting into the ABCs of soil horizons and how they characterize a soil series, let's take a look at the defining properties. After all, found in a multitude of combinations, they are the fundamental qualities that have yielded more than 475 distinct soil series in the state. Taken together, they are also what determines the suitability of different soils for diverse ecological habitats and human uses. My focus here will be on the physical and chemical characteristics. The biology and microbiology of soils are important enough to get their own chapter later in the book—see chapter 7.

The Architecture of Soils

One of the most obvious physical characteristics of a soil is its texture. Many people can at least identify a soil as sandy or clayey. Most people have also heard the term "loam," but how many really know what loam is? As a child growing up on the family potato farm and hearing my father talk about our loamy soils, I didn't understand exactly what that meant, and I had probably never even seen any other kind of soil. As it turns out, loam is great for growing potatoes, which could be one reason why my Polish ancestors settled in northern Michigan in the first place, where the soils and the crops suited to them were very similar to those of the glaciated landscapes of their homeland in west-central Poland.

The term "loam" describes one type of soil texture. Texture, which is probably the most influential of all the soil properties, refers to the relative amounts of sand, silt, and clay that make up a soil. For this purpose, sand, silt, and clay are defined only by the size of their individual grains, regardless of what mineral they're made of. An individual silt grain is smaller than a twentieth of a millimeter—typically impossible to see with the naked eye—and the largest individual clay particle is twenty-five times smaller than the tiniest sand grain. But what we can't see we can usually feel with our sensitive fingertips, so what soil scientists call the feel method is probably their most important skill, although it takes a lot of experience to develop great accuracy.

A soil made up mostly of silt has the silky feel of flour when rubbed between your fingers, while a sandy soil feels gritty. When I'd been working mainly with sandy soils in certain parts of Iowa, which left my hands feeling like sandpaper, I always welcomed the soft, silky feel of a silty soil. It is the preferred soil of gardeners and farmers everywhere. On the other hand, a soil with a lot of clay might feel as smooth as toothpaste when wet but cloddy and hard as a rock when dry. Like a child's Silly Putty or the exercise putty from your physical therapist, moist clay sticks together and can be easily molded into shapes. This is because it exists as aggregates of tiny platelets held together by chemical forces rather than as separate grains. Individual clay platelets, about a fiftieth the thickness of a sheet of copier paper, can be seen only with an expensive electron microscope.

Tama soils contain about 65 percent silt. In fact, soft pleasing silt is the dominant component in the majority of Iowa's soils because they formed in the thick deposits of loess that fell during the last ice age. In chapter 3, we'll look in more detail at that event and other geologic episodes that shaped Iowa's land and largely predetermined its present-day soils. Texture is controlled mainly by the nature of the geologic material in which a soil formed, called its parent material. For this reason, texture is also the most stable soil characteristic and can remain basically the same over centuries and even millennia, especially in the lower part of the soil profile.

The size and shape of individual sand, silt, or clay particles have a big influence on how they get packed together, which determines the soil's porosity—the amount of empty space between particles that holds air or water. In turn, soil texture and porosity have a major impact on how a soil behaves and responds to various inputs and conditions. For example, how fast rainwater can percolate through a soil profile and how much water that soil can store for plant use differ tremendously depending on its texture. Water moves through the relatively large pores of a sandy soil much faster than it can percolate through a clayey soil, but a soil rich in clay holds more water in its abundant tiny pores much longer, so it doesn't dry out as quickly.

This difference in porosity greatly affects plant growth—for example, the ease with which plant roots can penetrate the soil and the amount of water available for growth. It also has a major impact on land use. For farmers, it is a key factor determining the ease of tillage, the effects of drought or excessive rainfall, and the likelihood of runoff and erosion. For homeowners, business owners, and municipalities, texture and porosity affect building foundations and basements and dictate the suitability of locations for landfills, septic tanks, landscaping, and the like.

Like all other natural features of our world—rocks and minerals, plants, animals, insects, even bacteria—soils have been classified into different categories for a long time. One way of classifying soils is on the basis of soil texture. The main textural classification scheme used in the United States is illustrated by an equilateral triangle, which the U.S. Department of Agriculture first developed in 1927 and has modified a number of times since (fig. 4). Each corner of the triangle represents 100 percent sand, silt, or clay. The triangle represents all the possible combinations of these three components and divides them into twelve classes of soil. As you can see, six of them are named some

4. USDA soil textural triangle. Courtesy of the National Soil Survey Center, *Field Book for Describing and Sampling Soils.*

kind of loam. Tama soils fit into the silt loam and silty clay loam classes in the lower right part of the triangle.

The term "loam" implies some mixture of sand, silt, and clay that exhibits the properties of all three about equally, even if the three are not present in equal amounts. For example, a soil near the bottom of the loam area on the triangle might have less than 10 percent clay but still have the properties of clay, such as the ability to be molded into a ribbon. Why is that? It's because a small percentage of clay has a much bigger influence on soil behavior than silt or sand. These tiny clay particles pack quite a punch. In fact, the type and amount of clay in the parent material strongly influence the kind of soil that develops.

Where I worked in the U.S. Southwest, most soils fell into the sandy zone in the bottom left portion of the triangle. Given the arid climate, they were mostly suitable only for growing cactus and mesquite, unless irrigated. The soils on our Century Farm in northern Michigan were mainly loam and sandy loam, great for growing potatoes and other tuber and root vegetables, because they have a texture that drains well but not excessively so. Likewise, the majority of soils in the north-central part of Iowa formed mainly in glacial deposits and are generally loam, sandy loam, or clay loam. However, Iowa's many soils that formed in loess contain very little sand, so they fall into the silt loam and silty clay loam classes in the bottom right of the triangle. They are also well drained and so are ideal for crops, pastures, and other grasslands. Tama soils, which contain less than 5 percent sand and are silt loam in some places and silty clay loam in others, are a perfect example.

A recurring motif in this book will be the striking clarity of physical soil properties as seen through a petrographic microscope in paper-thin slices of the soil, called thin sections. Observing soil under the microscope gives you a real sense of what the eminent Swiss American soil scientist Hans Jenny meant when he said of micromorphology that "the soil resembles abstract art . . . and speaks to us through the colors and sculptures of its profile, thereby revealing its personality" (Stuart 1984). Under a microscope, the complex architecture of a soil *as a system* is truly revealed.

A petrographic microscope is called that because a subgroup of geologists called petrologists use it to study thin sections cut from rock. To create thin sections of soil, a block of soil is first infused with a resin. Once it hardens, a thin slab is cut with a rock saw, glued to a glass slide, and abraded to the proper thickness. The photomicrographs in figure 5 compare two very different parent materials. The large grains in the loam (left) are mainly quartz with a few feldspar minerals. Some of the colors are false colors produced under the polarized light of the microscope; the slightest difference in thickness of a thin section will render different colors.

Lab tests showed this particular loam to consist of 40 percent sand, 35 percent silt, and 25 percent clay by weight. In contrast, the silt loam

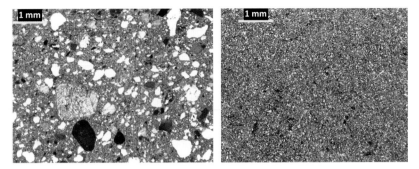

5. Parent materials with loam texture (left) and silt loam texture (right).
Photographs by the author.

(right) contains only 2 percent sand along with 71 percent silt and 27 percent clay, so you're able to see at most ten sand grains, particles larger than about a twentieth of a millimeter. This silt loam sample is from a loess layer, where quartz and feldspar dominate just as they do in the loam. But in loess, glacial ice had ground the mineral particles to a very fine size before the wind picked them up and redeposited them. Although glaciers aren't the only agent in nature able to grind quartz that fine, they are by far the most efficient, and ice is responsible for nearly all the thick deposits of loess in Iowa (Muhs and Bettis 2003). Chapter 3 contains more detail about the geologic deposits of the state and how they govern the texture of Iowa's soils.

Texture isn't the only property that controls the rate at which water moves through a soil and the weathering of minerals that results. Whereas texture is inherited from the geologic deposit in which the soil developed, soil structure develops over time as part of the soil-forming process, called pedogenesis. This word translates as the creation or origin of that which is underfoot, from the Latin *ped* meaning "foot"—think of the word "pedestrian"—and it refers to the slow development of soil features over time. The rate and degree of pedogenesis or soil formation depend on a number of factors besides the texture of the geologic parent material, as we'll see later.

Soil structure is the arrangement of particles into aggregates called peds, a shorthand reference to their origin through pedogenesis. The

Granular	Blocky	
(Soil aggregates)	(Subangular)	(Angular)
Platy		
	Prismatic	Columnar
Wedge		
Single Grain	Massive	
(Mineral/rock grains)	(Continuous, unconsolidated mass)	

6. Types of soil structure. Courtesy of the National Soil Survey Center, *Field Book for Describing and Sampling Soils.*

kinds of structure a soil exhibits vary with its location in the soil profile (fig. 6). The Tama topsoil exhibits granular structure. Granular structure forms mainly through biological processes, such as soil moving through the digestive system of earthworms. In figure 7, in a view under the microscope of granular structure in a healthy topsoil from Mahaska County, the granules are roughly a millimeter in diameter.

Blocky structure and prismatic structure are characteristic of the subsoil—the part of the profile below the topsoil—and are mainly the result of repeated episodes of wetting and drying or freezing and thaw-

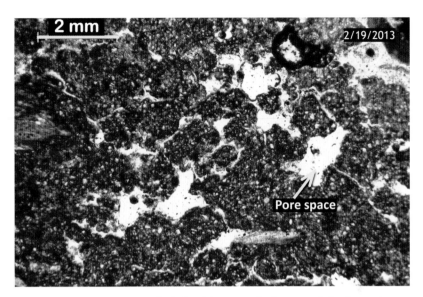

7. Granular structure typical of a healthy topsoil. Photograph by the author.

ing, both of which cause the soil to first swell and then shrink. Pore spaces between the peds of a soil showing a very well-developed blocky structure are the result of repeated swelling and shrinking and the persistent penetration of roots (fig. 8). The pores, generally greater than about a tenth of a millimeter wide, are called macropores. They provide room for air and water, which are necessary to nourish the soil microbes and roots essential for healthy plant growth, as discussed in chapter 7. Much smaller micropores existing *within* the soil peds are the home for most bacteria, which, contrary to some people's misconception of all bacteria as germs, are a very beneficial component of soil life.

The arrangement of peds and pores forms the framework for the architecture of a soil, in which the peds—the bricks—are held together by physical and chemical binding agents—the mortar—including clay particles, organic matter, and plant roots. In the mid-1990s, a U.S. Department of Agriculture soil microbiologist discovered another key binding agent: glycoproteins produced by certain fungi, which are present in healthy soils but largely absent from overworked soils. Under

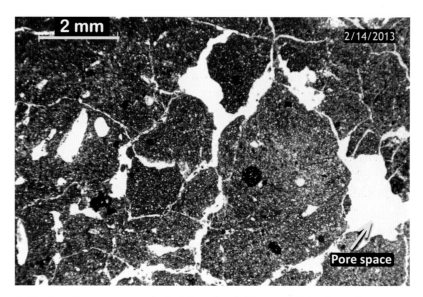

8. Blocky soil structure typical of the subsoil. Photograph by the author.

undisturbed conditions, these binding agents are present in abundance and soil structure becomes more pronounced over time. However, the disturbance caused by tillage and compaction by heavy farm machinery easily destroys soil structure. In Iowa, loss of structure is a major cause of reduced infiltration of rainwater and increased runoff from crop fields, with erosion as the result, as we'll see in chapter 6.

The most noticeable property of a soil is its color. After all, it was probably the first thing you noticed about the Tama soil profile pictured earlier. Soil color is the product of a number of processes and environmental factors and thus can provide clues about such things as soil fertility or the fluctuation or absence of a water table. The abundance and behavior of water in a soil are the principal factors influencing soil color, so climate is obviously key. A virtual rainbow of soil colors can be found across the United States, reflecting climates ranging from tropical to arid.

Iowa's climate of today doesn't vary enough across the state to produce such marked color differences. But there is another agent that

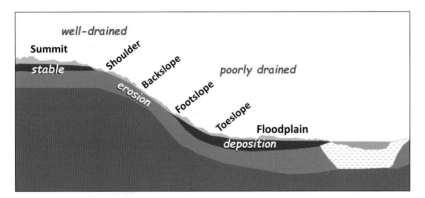

9. Components and conditions along a typical Iowa hillslope according to Robert Ruhe's model. Illustration by the author.

strongly affects color—one that Iowa has plenty of—and that is topographic relief. Soils that form in close proximity under the very same climate can exist at different levels or positions on the landscape, for instance, on hilltops and in valleys. Robert Ruhe, a distinguished and somewhat eccentric geomorphologist who worked for the Soil Conservation Service in Iowa in the 1950s and 1960s, authored the 1969 book *Quaternary Landscapes in Iowa*. His studies in southwest Iowa led him to propose a model in 1975 for the various slope positions typically present on the landscape, which earth scientists in the U.S. have used widely ever since. The most extreme difference in soil color often occurs between Ruhe's shoulder position, where erosion is common and water is retained very briefly, and the toeslope position, which receives the dark surface soil transported from above and where the water table is usually quite close to the surface (fig. 9).

In most Iowa soils, differences in color are due to the amount of organic matter that is present, which coats mineral grains and renders the soil black, and the oxidation states of the iron and manganese minerals that are present. Because the amount of water in a soil is very dependent on landscape position, so are the oxidation states of iron and manganese. To understand oxidation, think of your car as it ages. As the paint wears off, iron in the steel parts of the body becomes exposed to water from rain and humidity and joins with oxygen to form iron oxides, commonly known as rust. Likewise, soil color is a reliable indicator of moisture conditions and drainage and, thus, the degree of natural aeration. The shoulder position of a slope seldom stays wet for very long, so the soil there is oxidized and usually some shade of brown, while soils in the toeslope position are often wet and are black or dark gray from organic matter, sometimes to a depth of a few feet. On the other hand, soil on a broad, flat summit may be black in the top several inches and brown with reddish mottles in the lower part of the profile. However, because water on these flat areas doesn't run off but can only percolate downward, colors in the lowest part of the profile, below the water table, are often gray reduced hues rather than the brown hues of oxidized soil.

Color can be misleading, though. In some cases, it is not the result of soil-forming processes but instead is inherited from the original parent material. A perfect example of this is the Gosport soil series in southeast Iowa, which formed in 300-million-year-old Pennsylvanian shale and mudstone previously uncovered by erosion. Gosport soils inherited the color of the rock, a purplish color not seen in soils elsewhere in the state. In most Iowa soils, however, color is a significant and telling property. The color bible of earth scientists is the *Munsell Soil Color Book*, consisting of color charts first developed by Albert Munsell in 1905 to explain color to his art students. Descriptions of soil profiles contain very precise terminology based on hue (tint), chroma (strength of color), and value (lightness or darkness). Colors are notated by codes from the Munsell charts, like 10YR 4/6 (dark yellowish brown) and 5Y 5/2 (olive gray), as shown in figure 10.

The Munsell book of color charts contains at least a dozen hues—one chart per hue—and at least 440 color chips. They range from dark red to olive green to bluish gray and every gradation in between. Iowa's soils typically display various shades of black, gray, brown, and yellowish brown. Mottles caused by reduction and oxidation of iron and manganese below the topsoil are much more colorful, displaying six or seven different hues ranging from yellowish red to greenish gray. Soils in Iowa's deep loess deposits often display dark yellowish brown iron

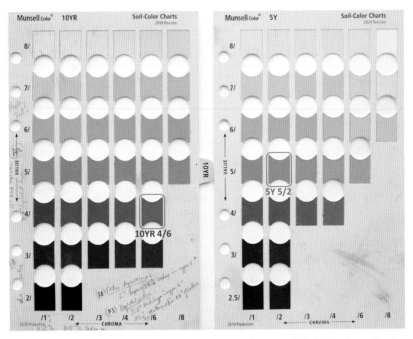

10. Pages for the hues 10YR and 5Y in the author's Munsell field book. 10YR 4/6 is dark yellowish brown and 5Y 5/2 is olive gray. From the *Munsell Soil Color Charts*, copyrighted and trademarked, Grand Rapids, Mich.

oxide concentrations that formed along root channels, as seen in the drilled core split in half in the lab in figure 11.

The Tama profile shown earlier has colors of black or very dark brown in the upper part and yellowish brown and grayish brown in the subsoil. Generally, the moist color of the topsoil is a good guide to the amount of organic matter in the soil. In general, black soils have more than 3.5 percent, dark brown soils have 2 to 3 percent, and yellowish brown soils have less than 1 to 2 percent (Magdoff and Van Es 2009).

Subtle differences in color can be significant. The color 7.5YR 5/4, which is called simply brown, can have an interesting pedigree. There are at least eight different chips called brown spread across the various hues in the Munsell book, not to mention light brown, pale brown, dark brown, and strong brown. As a geologist, I like the 7.5YR 5/4

11. Large tubular iron oxide features in loess. Photograph by Phil Kerr, Iowa Geological Survey.

version of brown because it is typical of a very old buried loess found in western Iowa called the Loveland Loess, which fell to the ground more than 150,000 years ago. It usually has a subtle but distinctive pinkish cast to it when compared to the very common yellowish brown (10YR 5/4) of many well-drained Iowa subsoils formed in younger loess.

At times, choosing the best soil color is like trying to find a paint color to match your wall. Despite the hundreds of Munsell color chips, it's not very often that you can find a perfect match, a testament to the extraordinary diversity of nature. It's also interesting how often people can disagree about the exact color. Explanations abound for this discrepancy: the lighting, the age of the person doing the describing, even the amount of caffeine in a person's bloodstream, according to one colleague.

An important soil property for farmers and engineers is density, which can significantly affect both plant productivity and drainage.

Soil density is measured as bulk density, which is the mass of dry soil for some unit of volume—such as grams per cubic centimeter or pounds per cubic foot—and is directly related to how porous the soil is. Bulk density depends first on the parent material and second on land use and management. For example, a dense glacial till, deposited and compacted by several thousand feet of overriding ice, might have a bulk density on the order of 2 grams per cubic centimeter. Put into practical terms, a 5-gallon bucket of that material once you dried it could weigh more than 85 pounds. A bucket filled with an *uncultivated* loess soil such as Tama silt loam would weigh only about 45 pounds once it was dried, but the same volume of a *cultivated* Tama topsoil typically weighs 50 to 55 pounds. If compacted over the years by repeated passages of heavy farm machinery, that same dried soil could weigh as much as 65 pounds.

Another physical property of soil related to density is its consistence. We describe moist soils as loose, friable, or firm, indicating how easily a soil can be crushed between thumb and finger. Consistence can also refer to how plastic or moldable the soil is, which is determined by rolling moist soil between your hands to form a string or by forming a ribbon between thumb and forefinger. At a modest water content, the length of the string or ribbon that can be formed before it breaks gives an approximate idea of the soil's clay content. This is a simple but important field test for identifying whether a loess soil, for example, is a silt loam or has enough clay to be a silty clay loam. Soil scientists with years of field experience have impressively discerning fingers and can identify clay percentages that are accurate within 1 to 2 percent.

Several of the properties described so far are quite dynamic in nature, particularly in the topsoil of cultivated soils. Chief among these are the amount of organic carbon in the topsoil and, as a result, color; the bulk density, which influences porosity; and the stability of the peds that make up the soil structure. These and others have recently been designated the dynamic soil properties, because they can change with land use and disturbance over the human timescale—in other words, in decades to centuries. Since these properties greatly influence soil fertility, the rate at which water can infiltrate the soil, and several other soil functions, soil scientists have been studying them intensively in recent years.

The ABCs of Soil

The vast majority of Iowa's soils formed in unconsolidated geologic materials deposited by ice, water, wind, or gravity. Some soils in northeast Iowa and a few other places formed in the loose residual layer at the top of bedrock, which centuries or millennia of weathering and disintegration of rock at the earth's surface had yielded. And new soils are continually forming in materials that erosion or the earth-stripping activities of modern humans expose to the air. In their original state, none of these geologic materials can really be considered soil in the sense of an ecosystem composed of mineral matter, organic matter, and living organisms that serves as a medium for plant growth. You could even get away with calling them dirt. And if you're wondering what the difference is, you can go to YouTube and watch an entertaining four-minute video called "Don't Treat It Like Dirt" by some creative high school kids from Panora, Iowa.

Soil formation begins soon after geologic materials are exposed to the atmosphere, and it starts at the top where water, air, and living organisms can most easily infiltrate. This may seem obvious to some, but geologists like myself are trained to think in terms of things happening from the bottom up; that is, older layers, events, and fossilized life-forms get covered up by consecutively younger ones. So thinking top down can be a bit of a mental shift for geologists contemplating the development of soil. It should also lay to rest the common misconception that soil is laid down as a deposit.

With sufficient time, a geologic deposit undergoes changes at various levels, eventually forming soil horizons. Soil horizons are horizontal zones of earth material that have gradually come to exhibit properties—physical, chemical, and biological—different from the parent material in which they formed. These differences develop as a result of four main processes that often occur simultaneously in one part of the soil or another: something is lost, something is gained,

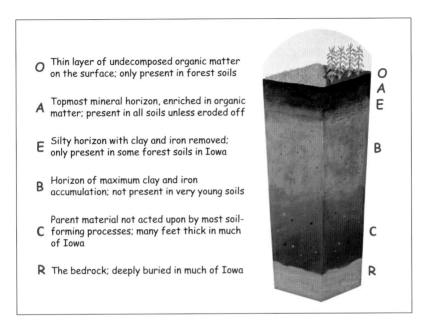

O — Thin layer of undecomposed organic matter on the surface; only present in forest soils

A — Topmost mineral horizon, enriched in organic matter; present in all soils unless eroded off

E — Silty horizon with clay and iron removed; only present in some forest soils in Iowa

B — Horizon of maximum clay and iron accumulation; not present in very young soils

C — Parent material not acted upon by most soil-forming processes; many feet thick in much of Iowa

R — The bedrock; deeply buried in much of Iowa

12. Composite representation of an Iowa soil profile. Illustration by Rebecca Kemble.

something is moved from one level to another, and some substances are transformed into other substances (Simonson 1959). The branch of soil science concerned with how soils form is called pedology, which also includes soil classification. The sequence and thickness of various horizons form the foundation for the classification of soils.

The soil profile is divided into major horizons, now known as the O, A, E, B, and C horizons (fig. 12). Where consolidated bedrock is fairly close to the surface, it is often called the R horizon, although it is not technically part of the soil. Two Russian scientists first introduced the concept of the major soil horizons in 1893 at the Chicago International Exposition, when they exhibited a display of mounted soil monoliths shipped by boat—large vertical columns of undisturbed soil to a depth of 3 to 4 feet (Gardner 1957).

In most Iowa soils, the first layer or horizon to form at the ground surface is the A horizon, and the presence of just a thin A horizon is sufficient for something to be called a soil. Informally referred to as the topsoil, A horizons in Iowa are normally made up of 90 to 99 percent mineral matter, including clay-sized to sand-sized grains of quartz, feldspar, mica, and many other minerals in smaller quantities. The remainder consists of decomposed organic matter, plant and animal fragments, fecal pellets of mites or earthworms or soil insects like springtails and beetles, seeds, pollen, other organic particles, and living organisms. The organic matter, of course, is what gives the topsoil its fertility. In Iowa, the organic matter in cultivated topsoil can range from more than 7 percent to less than 1 percent, about 60 percent of which is organic carbon. The A horizon can be thick, up to 18 inches or more in native prairie soils, but it is generally much thinner in cultivated soils. Today, the A horizon in Tama soils varies from about 10 to 14 inches thick, although it was undoubtedly thicker prior to cultivation. Loess soils under cultivation, such as Tama soils, are easily eroded when they are not protected by conservation efforts.

In prairies, much of the organic matter comes from dead vegetation and decomposing roots, but in forests it mainly works its way down from an overlying horizon of decomposing leaf litter, woody debris, and moss. This top layer in forest soils and some woodland soils is designated the O horizon. It is the most dynamic horizon and may not even be present all year round because, given sufficient moisture and earthworm action, it can decompose and be incorporated into the A horizon below it in as little as a few months (Schaetzl and Thompson 2015).

Another major horizon is the E horizon, which is not commonplace in our state except in some soils of eastern Iowa that formed under forest vegetation. The *E* stands for "eluvial," which describes material leaving a soil layer. In contrast, the B horizon is considered to be an illuvial horizon, where the eluviated material accumulates. (The beginning *e* and *i* of these two words are analogous to the same letters and meanings found in "emigrate" and "immigrate.") The E horizon is usually light in color and silty because the clays, iron, and organic matter have been removed by the relatively acidic water percolating through the A and O horizons in humid environments, especially evergreen forests. The E horizon, when present, is almost always the thinnest horizon, usually just a few inches thick.

Apart from young soils and soils with an E horizon, most Iowa soils have a B horizon immediately below the A horizon. It is generally lighter in color than the A and O horizons, displays blockier structure, and often contains more clay and iron and aluminum compounds than the original parent material. The B horizon can be thought of as a zone of accumulation. As soil formation proceeds, minuscule clay particles in the A horizon become suspended in rainwater, which percolates deeper into the profile through soil pores between the peds. The clay comes out of suspension in the B horizon due to evaporation or a change in pH, the acidity or alkalinity of a soil. Other things like iron, manganese, and aluminum compounds are also dissolved and move downward in solution to the B horizon, where they come out of solution to form visible concentrations of colorful iron oxides. Although the B horizon is less fertile than the A horizon, it is important to plant growth because it has a big influence on water movement and root development.

The major soil horizons are often characterized further by the use of suffixes. The field guide for describing soils lists some thirty-five possible suffixes, an indication of just how complex soils can be (National Soil Survey Center 2012). At least eight of these can be found in Iowa. One ubiquitous example in our agricultural state is the Ap horizon, which is the plowed or disked part of the A horizon in cultivated fields. The B horizon has several possibilities. One of the most commonly seen is the Bt horizon, where the *t* indicates a significant accumulation of clay that moved down from the A and E horizons. The *t* comes from *ton*, the German word for clay. Bt horizons typically display coatings of clay on the surfaces of peds and channels, where clay platelets have been deposited parallel to the surfaces.

Such clay coatings are easily seen under the microscope and are sometimes very striking. In figure 13, you can see abundant clay coatings on worm and root channels in the Bt horizon of a Madison County soil formed in loess. I took the photomicrographs under two different kinds of light exposure on the petrographic microscope. Under plane polarized light (left), which is composed of light waves vibrating in only one direction, as with polarized sunglasses, the clay coatings are their natural brownish color and are difficult to see. However, they

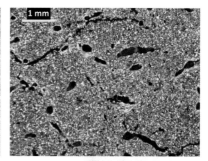

13. Clay coatings on channels in a Bt horizon under plane polarized light (left) and circularly polarized light (right). Photographs by the author.

show up brilliantly under circularly polarized light (right), which is composed of light waves vibrating in several directions 45 degrees apart. There the coatings display bright colors due to the double refraction of light by some clay minerals. Each coating consists of several very thin layers of clay that came out of suspension at different times, and it is the laminated orientation of the clay particles in the coating that increases the brightness of the colors.

Another type of B horizon is the Bk horizon, where the k indicates a concentration of calcium carbonate—the mineral calcite—brought in from elsewhere in the profile. (In case you're wondering, c had already been taken and the Russian word for calcium begins with k.) Calcium carbonate is the main component of limestone. Because most of Iowa rests on limestone at some depth in the rock column, drinking water across the state contains a fair amount of lime dissolved from the rock. Most Iowans are familiar with the scaly white deposits that form in water heaters and teakettles as water heats and evaporates. In the same way, as soil moisture evaporates, calcium carbonate precipitates and forms white concentrations, concretions, and nodules in pores and along fractures in the subsoil.

Figure 14 shows a soil pit in Adams County, with numerous white calcium carbonate nodules exposed in the thick Bk horizon of a soil formed in glacial deposits. A microscopic view of one of the carbonate nodules reveals its tiny calcite crystals (fig. 15). Bk horizons are

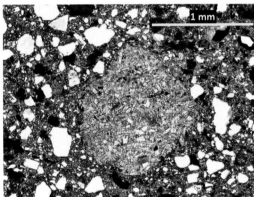

14. (*Above*) Soil pit in Adams County showing a Bk horizon. Photograph by Ben Woida Clark.

15. (*Left*) One of the calcium carbonate nodules from the soil pit in figure 14 as seen in a thin section. Photograph by the author.

common in Iowa because in their original state most of the parent materials for our soils contained finely dispersed calcium carbonate. As rainwater percolated through these materials, it slowly dissolved the calcium carbonate out of the A and upper B horizons and carried it downward to the lower horizons. This process is commonly known as leaching. The lime then precipitated out of suspension to form the characteristic white concentrations of the Bk horizon. If you've traveled in the arid American Southwest, you may have seen a more dramatic version of this—white rock-like caliche that began as a Bk horizon at the bottom of the evaporation zone. Over time, calcium carbonate cemented all the soil particles together and the zone grew to be many feet thick.

An interesting B horizon seen in Iowa is the Bss horizon, where the *ss* stands for "slickensides," which are glossy faces produced by shearing when one soil mass slides past another. During shearing, the clay is smeared to form a veneer that can be blindingly bright in sunlight when exposed. Large, well-developed slickensides are present in the same soil pit in figure 14. Lab tests done on this soil showed that the horizon at that depth had the texture of silty clay and contained about 45 percent clay—roughly midway up the texture triangle. Slickensides form only in soils containing a lot of clay, because when clay is wet it swells and causes pressure to build up. Soil at 4 feet below the surface, as in this example, is too deep to force its way up vertically, so horizontal shearing is the only option left.

The last major horizon in the soil alphabet, the C horizon, occurs below all the B horizons. In very young soils, it occurs immediately below the A horizon because not enough time has elapsed for a B horizon to form. The C horizon is relatively unaffected by biological activity and soil formation. It has a texture nearly identical to the soil's original parent material and has very weak or no soil structure, but it usually does show some evidence of mineral weathering. It is common to find prominent concentrations of orange iron oxides and black manganese oxides in the C horizon, often precipitated out along fractures.

The C horizon matrix generally is not leached of calcium carbonate like the B horizon and often has soft white masses or hard nodules that formed from the dissolved carbonate carried down from the B horizon. In loess soils, these lobed nodules are 1 to 4 inches long and often roughly resemble tiny dolls. In fact, geologists call them loess *kindchen* from the German word for small children.

The boundaries between soil horizons are sometimes sharp but more often are gradual. The thickness of the transition zone between one assemblage of soil properties and another is usually somewhere between 1 and 5 inches. A zone with characteristics of both the overlying and underlying horizons is often considered to be a transitional horizon and labeled, for example, as AB, BA, BC, and so on. The bottom of the C horizon is especially imprecise and somewhat open to interpretation. Where a soil formed in loose earth material weathered out of bedrock, like some soils in northeast Iowa, the bottom of the C horizon is where solid rock—the R horizon—begins. Of course, in most of Iowa glacial sediments are extremely thick and the depth to bedrock ranges from fewer than 10 feet to several hundred feet. Rather than lumping all this into a C horizon, some have proposed that this zone be called the D horizon, consisting of unaltered sediments that would not be considered part of the soil profile (Tandarich, Darmody, and Follmer 1994). This includes extremely dense glacial till that has remained submerged below the water table and unchanged since it was first deposited. When brought to the surface in cores, this very dark gray till provides a palpable encounter with the ancient past—it looks, feels, smells, and probably tastes exactly as it did when immense glacial ice hundreds of thousands of years ago lodged it onto the land surface.

CHAPTER TWO

Wealth in Diversity

I'm told there are people who do not care for maps,
and I find it hard to believe.
—ROBERT LOUIS STEVENSON, 1883

AS YOU'VE ALREADY SEEN, one relatively simple way of classifying soils is on the basis of texture, which is largely inherited from their parent materials. But soils are so much more than the geologic sediments in which they formed. They consist of geologic sediments influenced and often radically changed by agents or conditions in their environment over long periods of time. Perhaps more than anything else on Earth, soils are the substance of transformation.

There are multiple and complex agents acting on the landscape, and since the work of Hans Jenny in the 1940s, we have grouped these agents into five soil-forming factors: *parent material, climate, organisms, time,* and *topography* all acting over a period of time. Chapter 3 will illuminate how these agents have modified the various geologic deposits of Iowa over the past several thousand years. It is these soil-forming factors and their combinations and interactions that are responsible for the truly diverse nature of soils in Iowa and their classification into more than 475 soil series.

Humans have differentiated among soils since postnomadic civilization began as they determined which soils were most productive for growing food through a process of trial and error. Attention to soils was necessary for immediate survival, and some early societies even passed on knowledge of soils and farming methods in agricultural manuals (Schaetzl and Thompson 2015). The Chinese appear to have practiced rudimentary mapping and classification of soils as early as 200 BC. During the heyday of ancient Greece, both Plato and Aristotle wrote about the great loss of soil fertility and productivity around Athens due to severe erosion. Aristotle's student Theophrastus documented the existence in Greece of distinct types of soil composed of different layers. The organized study of soils as a discipline, however, is very young compared to the other natural sciences such as chemistry, biology, or even astronomy. Leonardo da Vinci was undoubtedly correct when he boasted in the sixteenth century that "we know more about the movement of celestial bodies than about the soil underfoot."

It wasn't until the nineteenth century that soil began to receive serious scientific attention, notably from Charles Darwin, who spent many years observing the activities of earthworms in the soil and devoted his last book to the topic—a book that sold better during his lifetime than his *On the Origin of Species*. The chemical analysis of agricultural soils was also evolving rapidly, although the findings often did not correlate well with crop production since scientists at the time assumed the results to be independent of physical properties such as texture. Around the same time, geologists began to formally classify soils—first on the basis of texture, then subdivided by agricultural productivity. E. W. Hilgard began mapping geology and soils for land-use planning and management in the southern U.S. around 1860. The 1880s saw the birth of modern soil classification when the pioneering geologist Vasily Dokuchaev published a work called *Russian Chernozem* and other texts, in which he classified loess soils of the great central Russian steppes according to their different profiles and properties. Dokuchaev, who is considered the father of modern pedology, was the first to note that these differences were related to both climate and vegetation patterns (Gerasimova 2005).

The contributions of his students later refined Dokuchaev's work. Most notable was Konstantin Glinka, whose work on soil groups of the world was translated into English by Curtis Marbut and became the foundation for the American system of soil classification (Schaetzl and Thompson 2015). Marbut was a geologist and geomorphologist from Missouri who worked for the U.S. Department of Agriculture and is considered the founder of American pedology. The early American system developed under Marbut paid tribute to the Russian groundwork through its names for several soil groups, such as Brunizems and Chernozems for two types of American prairie soils, which are very similar to soils on the Russian steppes. Marbut's system was used in Iowa for several years before the USDA formally published it in the 1938 *Yearbook of Agriculture* (Baldwin, Kellogg, and Thorp 1938).

Soil Classification in Iowa and the Nation

Soil classification in Iowa, which saw some of the very earliest soil mapping in the country, began in 1901 in the Dubuque area. Soil survey reports for the Dubuque area, Cerro Gordo County, and Story County appeared in 1903, followed by Tama County in 1904 (Fenton and Miller 1982). Starting in 1913 with Bremer County, soil surveys were conducted by county under the National Cooperative Soil Survey, which had been established in 1899 and continues as the umbrella for USDA soil-mapping activities to this day. The Soil Survey is a joint federal and state undertaking with a long history of success, symbolized by today's online database of soils and soil maps for all fifty states, the largest natural resources database in the world.

In 1936, when all but thirteen Iowa counties had been completely mapped, the Iowa Agricultural Experiment Station summarized the results in *Soils of Iowa* (Brown 1936). At that time, sixty soil series were identified across the state. Mapping in some of the remaining thirteen counties was not completed until after World War II, a time that saw budgetary and staffing limits on all but war-related efforts. Publication of county soil surveys, which had been issued at a rate of nearly one per year since 1914, stopped completely between 1941 and 1950. The last county survey to be published was for Allamakee County in 1958, although the mapping had begun nearly ten years earlier.

In the 1950s and 1960s, soil classification in the United States underwent major changes that led to the system currently in use in much of the world (Effland, Eswaran, and Helms 2005). The development of this new system also resulted in the eventual remapping of all of Iowa's counties. It is important to acknowledge the major role that soil scientists from Iowa played in this overhaul of U.S. soil classification. It all began in the early 1950s under Guy Smith, a native of Atlantic, Iowa, and a faculty member in Iowa State University's Department of Agronomy at the time. Gerald Miller, Iowa State professor emeritus of soil science, tells an interesting story about the beginnings of the new system. While walking from Curtis Hall to the Iowa State Memorial Union for lunch one day, Smith was bemoaning certain shortcomings of the existing system when his companion, agronomy professor Frank Riecken, light-heartedly asked, "Well, why don't you develop a new one?" That challenge began twenty-five years of joint work between Smith and other ISU and USDA soil scientists, including Riecken and Roy Simonson.

In 1952, Simonson, Riecken, and Smith published *Understanding Iowa Soils*, a precursor to concepts incorporated into the modern national system. By that time, both Simonson and Smith had left Iowa State to become assistant chief and principal soil correlator, respectively, for the USDA's Division of Soil Survey in Washington, D.C. Smith later served as director of the Soil Survey. They are only two of several distinguished names in the field of American soil science whose career paths wound through Iowa at one time or another. For example, Creston native Richard Arnold served as the national director of the Soil Survey for sixteen years. Their contributions pay tribute to the prominent role that the study of Iowa soils has played not only in the history of our own state but on the national stage as well.

At the national level, the new U.S. classification system that resulted from these roots was released for comment and criticism as a series of seven iterations and additional supplements over a period of some eighteen years. If, as one reviewer facetiously claimed, a hidden goal of classification is to provide something to revise, American soil science outdid itself (Handy 1964). The final preliminary document, released in 1960, was commonly known as the *Seventh Approximation* (which I've always thought would make an interesting title for a movie). Five years later, Agronomy Department faculty members, including Frank

Riecken, who had stayed at ISU, jointly authored *Principal Soils of Iowa*. It summarized the important changes in classification of Iowa's major soils that had occurred after about 1950 (Oschwald et al. 1965). It also established the principle that Iowa's soil series do not occur randomly on the landscape. Although out of date, this publication is still useful today for its clear synopsis of the state's principal soil association areas. A soil association is a repeating pattern of two or three major soil series occurring adjacent to one another within a geographic area. For example, one soil series might typically occur on the uplands, another on steep slopes; one might be found on terraces, another on footslopes or swales; and so forth. One example is the Shelby-Sharpsburg-Macksburg association in southwest Iowa. Because they occur at different slope positions, the soils in an association differ in their moisture characteristics and therefore in their physical properties.

The Soil Survey finally published the new system of classification, called *Soil Taxonomy*, in 1975. It dropped many of the Russian names for soil groups in an effort to distinguish soils by their physical and chemical properties rather than by inferred climatic and vegetative influences, as the Russian system had. (The modern Canadian system retains more of the Russian terminology than the American one.) When I began studying soils in 1987 for my Ph.D. research, the "big green book" was the soil classifier's bible. It went on to become the "big purple book" when the Soil Survey staff published a second edition in 1999. Each edition weighed about 6.5 pounds, but you can now read its 900 pages online in a weightless pdf format. It includes updates through 2004, with more recent updates available as separate monographs.

Soil Taxonomy is used in as many as fifty countries around the globe, some notable exceptions being Russia, China, Australia, Brazil, South Africa, and Canada (Krasilnikov and Arnold 2009a). In the distribution of soils around the world, *Soil Taxonomy* defines twelve main kinds of soil, called soil orders. The soil order is the highest level in the hierarchy of soil classification. The orders are subdivided into successively more and more specific categories, called suborders, great groups, subgroups, families, and finally the soil series. The subdivisions have things in common—for example, they all have mineral and organic components—but they may also have radical differences. To invoke a zoological analogy, vertebrates of the animal kingdom are similar in one major respect, but the diversity of organisms with backbones is tremendous, from swimming whales to the tiniest burrowing rodent. So it is with soils. For the sake of the lay reader, this book will emphasize only the first and last of the levels, the order and the series, but I will touch on a few distinguishing criteria for some of the suborders and great groups to more fully illustrate the wide variety of soils in Iowa.

The main soil orders in Iowa are Mollisols, Alfisols, Inceptisols, Entisols, and Histosols, which are prairie soils, woodland soils, young soils, very young soils, and organic soils (fig. 16). We also have a meager handful of Vertisols—clay-rich soils that develop deep cracks during dry seasons. The language enthusiast might find it interesting to know how these names came to be. While the Linnaean system of classification for living organisms draws almost exclusively on Latin for its taxonomic names—for example, Mammalia, Carnivora, Canidae, and *Vulpes* for foxes—the names given to soil orders and their subdivisions use a combination of syllables taken from Latin, Greek, French, and English. Note that each soil order name ends in "sol," which comes from the Latin *solum*, meaning "soil." The "moll" in Mollisol comes from the Latin *mollis*, meaning "soft," and refers to the fluffy texture of a prairie soil's thick topsoil. The "alf" in Alfisol refers to the symbols on the periodic table for the aluminum (Al) and iron (Fe) that often enrich a woodland soil's B horizon. Inceptisol comes directly from the Latin *inceptum*, meaning "beginning," and the syllable "ent" in Entisol probably came from the English word "recent." The "hist" in Histosol is from the Greek *histos*, meaning "tissue," while the "vert" in Vertisol comes from the Latin *verto*, meaning "to turn or mix."

Without naming all the subdivisions of the common soil orders found in Iowa, which would constitute a table several hundred pages long, it will be helpful to explain certain important criteria used to divide these soil orders into suborders. One very important influence on soil behavior, such as its fertility and drainage, is the moisture condition of the soil environment, which is reflected in the suborder name. For instance, a Mollisol that is saturated for significant periods of time

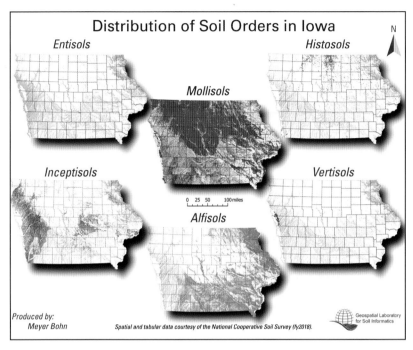

16. Distribution of soil orders in Iowa. Maps by Meyer Bohn.

during the year is called an Aquoll, where "aq" is obviously taken from *aqua*, the Latin word for water, and the ending "oll" is taken from the order name Mollisol. Aquolls are common in the prairie pothole region of north-central Iowa and in stream valleys across the state. However, the majority of Mollisols in the state occur higher on the landscape and have much better drainage. They are called Udolls (pronounced "yew-dolls" by soil scientists), where "ud" comes from *udus*, the Latin word for humid. Likewise, the most common Alfisols in Iowa belong to the Aqualf and Udalf suborders. Again, note the "alf" ending of the names, taken from the name of the Alfisol soil order.

A soil series is roughly equivalent to a species in biological taxonomy, and each soil series belongs to a taxonomic class. For instance, the Tama series belongs to the class Argiudolls. If you work the name Argiudoll backward, you can see that it is a Mollisol (prairie soil) in the suborder Udoll (under a humid climate) in the great group Argiudoll (with an argillic horizon, another name for a clay-rich Bt horizon). Thus, these taxonomic names communicate the order, suborder, and great group levels of the soil series in an immediate and elegant way.

Prior to the twentieth century, more than two-thirds of Iowa was carpeted with rich Mollisols. Regrettably, intensive agriculture has eroded some of them so severely in parts of the state that they no longer have a thick enough A horizon to qualify. Today, about 60 percent of Iowa's soils are technically Mollisols. Although very few of these superb soils exist in their original state, you can find an undisturbed Mollisol soil with a profile very similar to what it may have looked like 150 years ago, before Euro-American settlement, in the few prairie remnants scattered around the state, along old fencerows, and in a few pioneer cemeteries. I recently probed into the soil at Doolittle Prairie State Preserve south of Story City, a prairie remnant that has been under native vegetation since before settlement and land survey records, and found a rich black topsoil more than 20 inches thick. Remarkably, the topsoil was even thicker on a grassy summit in Hamilton County, in a pasture that had never been cultivated where a small dairy herd grazed intermittently under careful management.

Mollisols of our friend the Tama series grace more than 933,000 acres in the eastern half of Iowa ("Iowa State Soil: Tama Soil Series" 2020). The seventh most abundant of the state's 475-plus soil series, Iowa's state soil formed in loess parent material and is characterized first and foremost by a thick dark A horizon usually between 9 and 16 inches thick (see fig. 2, chap. 1). Even as deep as 3 feet down, Tama soils and other Mollisols may contain 1 percent organic carbon, which is at least ten times more than Iowa's other soil orders hold at that depth. The underlying B horizons in a Tama soil can be more than 3 feet thick. They include a Bt horizon with silty clay loam texture and blocky or sometimes prismatic structure. You can see an actual Tama soil by visiting the Science Center of Iowa's exhibit about the state's historic prairie, which includes a monolith of the Iowa state soil collected in Poweshiek County in 2019.

Alfisols cover 25 percent of Iowa's land surface, primarily in eastern and southern Iowa on the gently rolling to steep topography of the

17. Fayette soil profile, an Alfisol from Winneshiek County. Photograph by Jon Sandor.

major river valleys. The second most abundant soil series in the state, the Fayette series, is an Alfisol. Its 1,400,000 acres are concentrated in northeast and east-central Iowa but extend into central Iowa valleys as far west as Des Moines. Although Fayette soils formed in loess just like Tama soils, they have an A horizon only 2 to 4 inches thick because they developed under trees rather than deep-rooted grasses.

Beneath the thin A horizon of Alfisols, there may be a grayish E horizon with silt loam texture and platy structure, although many Fayette soils have lost their E horizons because of long-term cultivation (fig. 17). However, they usually have at least a transitional BE horizon, which has more clay than the A horizon above but much less than the Bt horizon below. An example of an Alfisol that still has its E horizon is

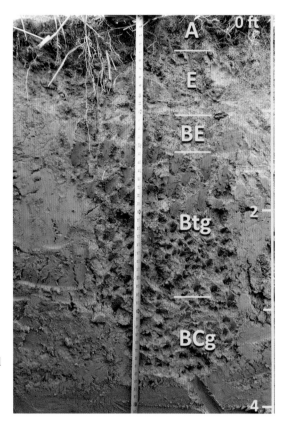

18. Coppock soil profile, an Alfisol from Appanoose County, with its light-colored silty E and BE horizons. The divots are tool marks. Photograph by the author.

the Coppock series found in the soil-judging pit in Appanoose County featured in chapter 1 (fig. 18; see also fig. 3).

Like most Alfisols, both the Fayette and Coppock soils have Bt horizons with blocky structure. Because Coppock soils are found on low, seasonally wet landscape positions as stream terraces and footslopes, their Bt horizons are often saturated and so exhibit grayish colors due to the reduction of iron. They belong to the Aqualf suborder. On the other hand, Fayette soils are found on upper landscape positions and have good drainage, so they belong to the Udalf suborder.

Entisols cover about 7 percent of the state's land surface. They are found primarily in two landscape settings: valley bottomlands and low

19. Nodaway soil
profile, an Entisol
from Pottawattamie
County. Photograph
by Jason Steele,
USDA–Natural
Resources Conserva-
tion Service.

terraces across most of the state and steep loess hillslopes in western
Iowa. These are dynamic landscape positions, where soil development
cannot keep pace with the deposition of sediment by streams in the
case of valley bottoms or erosion in the case of steep hillslopes. Entisols
have not had enough time to develop B horizons, so their profiles consist
simply of A and C horizons. Where they formed in alluvium—sediment
deposited by flowing water in streambeds and on floodplains—the
parent material is commonly stratified into layers with silty and often
sandy textures in their original state. In older soils, soil-forming processes have obliterated such stratification. A good example of an Entisol formed in stratified silty alluvium is the Nodaway series, a very
common soil in valleys in the southern half of Iowa (fig. 19). It is classified as a Udifluvent, in which "fluv" signifies formation in a fluvial, that
is, a floodplain setting.

20. (*Right*) Storden soil profile, an Inceptisol from Story County. Photograph by Gerald Miller.

Another 7 percent of the Iowa land surface displays soils belonging to the Inceptisol order. These are young soils like Entisols, but they are old enough to have developed B horizons, although not Bt horizons. They are found on steep hillslopes in western Iowa where Mollisols have undergone severe erosion, but they also occur in many other landscape settings in diverse parent materials—till, loess, and alluvium—with a wide range of textures. For example, the Storden series occurs exclusively in north-central Iowa, where it developed on ridges in young deposits from the last glaciation that have not yet been leached of calcium carbonate (fig. 20). So if you were to apply a drop of 10 percent hydrochloric acid to this soil, it would effervesce "strongly" to "violently," as described by soil scientists. Storden profiles consist of

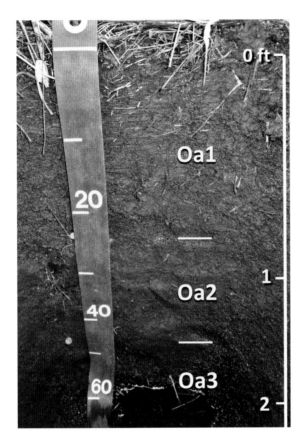

21. Houghton soil profile, a Histosol. Courtesy of the Michigan USDA–Natural Resources Conservation Service.

an A horizon over a Bk horizon, which is up to 4 feet thick and has fine blocky peds and white coatings of calcium carbonate. These soils are classified as Eutrudepts, in which "eutr" comes from the Greek prefix *eu*, meaning "good" or "fertile."

The majority of Histosols in Iowa occur in the north-central part of the state. While there are many Histosols to our north in Minnesota, only three series represent the order in Iowa, totaling about 77,000 acres. One of these is the Houghton series, which is concentrated in Cerro Gordo, Hancock, Winnebago, and Worth Counties but is mapped from central Minnesota to eastern Michigan and as far south as Indianapolis (fig. 21). Interestingly, the series was first established nearly a hundred years ago in the north-central Lower Peninsula of Michigan,

22. Prairie potholes in north-central Iowa. Photograph by Lynn Betts, USDA–Natural Resources Conservation Service.

about seventy miles from our family farm, in the area around a large glacial lake called Houghton Lake. The series type section is still in Michigan, which glaciers occupied even more recently than north-central Iowa.

In Iowa, Houghton soils are found in closed depressions called prairie potholes, on glacial moraines, till plains, former glacial lake plains, and outwash plains consisting of sand, gravel, and cobbles deposited by large glacial meltwater streams (fig. 22). They also occur on some floodplains around the state. They have a thick O (organic) horizon at the surface consisting of 4 to 6 feet of black to dark reddish brown muck—soil made up of highly decomposed organic matter with little fibrous material or identifiable plant parts. Houghton soils do not have a B horizon; the O horizon rests directly on the C horizon in glacial till. A Houghton soil is a Haplosaprist, "sapr" being derived from the Greek *sapros*, meaning "rotten," indicating this soil's highly decomposed organic matter. Farmers have drained and cultivated most Histosols in

23. Soil profile of a Texas Vertisol, in which extreme swelling of clays has produced microhighs and microlows from expansion as well as oblique shear planes and wedge-shaped peds below about 20 inches. Photograph by the author.

the prairie potholes of north-central Iowa for many years, so they have an Op horizon, a plow layer. Histosols on floodplains in the rest of the state have not been drained.

There are only two Vertisol soil series in Iowa, which formed in clay-rich alluvium on small areas all along the Missouri River, many too small to be shown on a map at the scale of figure 16. These are dynamic soils that typically contain more than 50 percent and as much as 75 percent clay, mostly of a variety with high potential for swelling. When the soils are wet, as is usually the case with these soils, the swelling generates shearing along oblique planes in the subsoil where slickensides typically form. As the soil slowly turns from the lateral stresses, low mounds and depressions—microhighs and microlows—form on the surface, as shown in the Vertisol from Texas in figure 23. During periods of drought, cracks 4 to 5 feet deep may develop. One property owner in Monona County told me his father used to joke that there were about fifteen minutes a year when their 90 acres of Vertisols of the Luton series were suitable for tillage.

Financial Implications

In this quick tour of Iowa's soil orders, I've given only brief attention to most of the levels in *Soil Taxonomy* in order to spare the reader. In short, soil classification has a very extensive, informative, and systematic vocabulary that is extremely helpful to informed users. But the classification of soils on cultivated land in Iowa is of much more than purely scientific or academic interest. For instance, whether a farmer has a hundred acres of the well-drained Sharpsburg series, which is a Typic Argiudoll, or a hundred acres of the somewhat poorly drained Lamoni series, an Aquertic Argiudoll, can mean a big difference in the assessed value of the land and therefore in annual property taxes. That's because Sharpsburg soils are capable of much higher crop yields than Lamoni soils. In the early 1970s, Iowa State professor Thomas Fenton developed something called the Iowa Corn Suitability Rating to capture these differences in yield potential. Most states in the Midwest have a similar rating, called a crop productivity index or a soil productivity index. Such ratings were not a new, or a midwestern, or even an American idea. Soil productivity ratings are used in many countries today, and even in the 1880s Vasily Dokuchaev interpreted his classification of soils in one region of Russia in terms of tax value assessment.

The Corn Suitability Rating was applied to Iowa soils for nearly forty-five years until Iowa State professor Lee Burras headed up the effort to modify the tool for greater compatibility with the revised *Soil Taxonomy*. Called the CSR2, it was adopted around 2015 (Burras and Owen 2012). Corn Suitability Ratings on cropland acres help farmers set cash rental rates and compute the sale value of farmland (Schwickerath 2015). They also provide a basis for counties to assess the productivity and value of farmland. Sharpsburg soils that formed in loess have a CSR2 rating of about 90 (out of 100), while Lamoni soils, which formed in a few inches of loess overlying an ancient clay-rich soil called a paleosol, might have a rating as low as 10. As a result, where these soil series are planted in row crops, the county assessor would assign a land value to the Sharpsburg acres that is eight to nine times higher than the value of the Lamoni acres. Not surprisingly, most of the questions soil scientists get from landowners pertain to the CSR2 values assigned to their agricultural land. Higher CSR2 values mean higher land values and higher property taxes, which can be a good thing or a bad thing depending on a landowner's plans for the land. In 2016, Iowa cropland was selling for $5,000 to $12,000 per acre. (The majority of settlers coming to Iowa in the mid-1800s bought land from the federal government for $1.25 per acre!)

Average CSR2s exist for every county in the state. They range from 39 in Decatur County in south-central Iowa, where slopes are steep, the loess cover is very thin or nonexistent, and many soils developed in clay-rich till or paleosols, to 92 in O'Brien County in northwest Iowa, where the loess is relatively thick and the land is level. The cash rental rates that farm operators pay to landowners strongly reflect productivity, along with several other factors such as field access, local grain prices, and longevity of the lease. In 2018, the average cash rent paid per tillable acre for corn and soybeans was $237 in northwest Iowa and $244 in northeast Iowa, compared to $174 in south-central Iowa. For an owner renting out 500 acres, that could mean a difference of up to $35,000 each year. So if an area is remapped to a different soil series, as occasionally happens, this can have substantial ramifications.

Mapping the Soils of Iowa

In addition to influencing land values, assigning a soil to a particular soil series makes available to the public a large amount of information on which land use and environmental decisions can be based. Each county in the state has a comprehensive document called a soil survey. In addition to detailed maps of the soil series in every part of the county, these surveys contain multiple tables with information on every soil's degree of suitability for such things as construction materials, building sites, sanitary facilities, and recreational development along with potential yields for crops, pastures, tree plantings, and the like. Soil maps and the accompanying tables are extremely useful, and farmers, private geotechnical and environmental firms, and state, county, and city planners rely on them extensively. It is surprising how many people are unaware of this extraordinary product, which the federal government provided for free in book form to the public from the 1960s until 2005 and online since then. This is one symptom of modern civilization's soil blindness—the lack of interest in the ground beneath our feet, the ultimate source of most of our food, fiber, and shelter.

Soil scientists remapped all ninety-nine of Iowa's counties between 1960 and 1989 and later updated roughly a third of the surveys again, with the last update completed in Worth County in 2014. The oldest surveys still in use date back to 1974, but since the advent of geographic information systems and computer mapping, the state's soil surveys undergo minor updates on a continuing basis. The National Soil Survey Center refreshes the official U.S. soils database annually and adds new interpretations to every soil to address up-to-date management concerns. For example, the data refresh in 2019 added a rating for suitability of a soil for aerobic soil organisms critical for soil health, a rating for susceptibility to compaction, and several others.

In the last several years, the focus has shifted from mapping soils to correlating soil series across county lines and addressing other inconsistencies in the enormous database. This came about around 2005 when the Natural Resources Conservation Service began conceptualizing soils on the basis of physical landscape characteristics—with the country divided into Major Land Resource Areas—rather than by political entities like counties. As a result, it became important to have consistency across the counties within a given MLRA. The agency called the correlation effort the Soil Data Join Recorrelation project. (Some folks with acronym fatigue thought it was easier to remember SDJR by the moniker Sammy Davis Jr.) Iowa is split into five MLRAs based on geology and topography, three of which overlap into adjacent states.

The huge undertaking of mapping and classifying the soils of the U.S. in the twentieth century involved scores of people—from university professors and soil scientists to cartographers, lab technicians, and data entry people who juggled enormous amounts of field and laboratory data, much of it before the age of computers. Some of the most passionate and determined folks in the world are the soil scientists who walked thousands of miles and probed literally millions of soil cores across this country, one county at a time. Every few years, they moved to a different county to map, sometimes to different states, and they became intimately familiar with the myriad soil series in their mapping area, with all their glaring or nuanced differences. Soils are like snowflakes in that no two are exactly alike, and it was the formidable task of the mapper to decide when to lump and when to split, with the goal of creating a map that was both accurate and practical for multiple uses.

Between 1901 and 2012, some 330 soil scientists participated in the field mapping and remapping of Iowa's soils, and dozens of cartographers and technicians labored to turn the field data into geographically accurate and user-friendly maps. Initially, they included men and women employed by the federal government, Iowa State University, and certain counties, which constituted a truly cooperative workforce. In later years, the U.S. Department of Agriculture employed all the soil mappers, but the important and massive job of cartographic output, map correlation, review, and quality assurance remained a joint effort on the part of the Natural Resources Conservation Service, Iowa State University, the ISU Extension Service, the Iowa Division of Soil Conservation, and numerous soil and water conservation districts. Laboratory testing of soil samples was done by ISU and later by the USDA's National Soil Survey Laboratory in Lincoln, Nebraska, where it continues today.

Just like the original mapping project, remapping Iowa's 36 million acres of soils using the classification system of the *Seventh Approximation* and *Soil Taxonomy* was accomplished by a hardy bunch of soil scientists. One of these was a young man from the Sand Hills of Nebraska, soft-spoken and of deceptively slight build, named Robin Wisner. He had grown up on the family cattle farm near the North Platte River, where at age eight he accepted a challenge from his grandfather to grow 4 acres of melons by himself, no small task in semiarid western Nebraska. In the 1960s, when he heard that the Soil Conservation

Service in Iowa was going to be hiring a few dozen soil scientists, he seized the wave of the future and decided to attend the University of Nebraska in Lincoln to study soil science.

Robin began mapping soils for the Soil Conservation Service in 1968 in Sac County, using an early draft of *Soil Taxonomy*, and didn't stop until 2014, after he had mapped in twelve Iowa counties. In addition, he served on months-long details in North Dakota in the latter half of his career. In each of these, Robin was the leader for a three-man mapping party, trained numerous young mappers on the job, and walked an average of 8 miles a day every weekday of the year when the ground wasn't frozen. Each county took three to five years to complete, which translates to 30,000 to 45,000 acres mapped each year. Mapping was accomplished mainly by pushing a three-quarter-inch soil probe 60 inches, later 80 inches, into the ground at least forty-five to fifty times a day, pulling it back out—the toughest part!—and drawing lines on Mylar placed over aerial photographs obtained by the USDA's Farm Service Agency every year or two. Now that he's in his seventies, Robin's lower back remembers those days well.

During my interviews for this book, I came to realize that hard work and personal initiative were characteristic of the folks who mapped Iowa, not surprising since most came from farming backgrounds. For example, Mark Minger, who grew up on a diversified farm in Dubuque County, began his own business at age sixteen raising registered Holstein dairy cows along with hogs in order to save money for college. He began college in Platteville, Wisconsin, taking courses in technical agriculture and working part-time, but in 1968 a connection with ISU's Department of Agronomy turned into an immediate opportunity to begin mapping soils in Black Hawk County while taking classes in Ames. In 1972, the Vietnam War interrupted his career, but even while serving overseas he managed to take an ISU correspondence course in physical geography to expand his understanding of soils.

When soil scientists finished mapping in one Iowa county, they moved to another county that often presented very different terrain and soils. After returning from Vietnam, Mark began his career with the USDA, first mapping in Buchanan and then in Audubon County. While he had mapped more than a hundred soil series in Black Hawk County, where the geology was complex, there were fewer than thirty series in Audubon County, where the geology of the upper several feet consisted almost exclusively of loess. By the time Mark retired in 2005, he had mapped for more than a dozen Iowa soil surveys and had worked in eighty-seven of Iowa's ninety-nine counties on other soil investigations: for the state's rural water pipelines, Department of Natural Resources woodland projects, wetland determinations, archaeological sites, highway construction, and others. His memories of extreme weather, barbed wire, venomous snakes, stinging insects, and threatening bulls are no doubt shared by all those who spent years walking Iowa's landscapes. On the other hand, experiences such as stumbling across Native American artifacts, finding layers of Yellowstone volcanic ash in the loess deposits of western Iowa, and simply being part of a massive mapping project that set the standard for many other states more than made up for those challenges.

Robin Wisner was one of at least eleven Iowa soil scientists who mapped more than a million acres from the 1960s on, including John Nixon, Maynard Koppen, Kermit Voy, Russ Buckner, Wilbur Jury, Melvin Brown, Wayne Dankert, Elton King, Bob Russell, Bob Vobora, and possibly a few others. Mark Minger, Chuck Fisher, Bob Dideriksen, Doug Oelmann, Herb Wilson, Mark La Van, Richard Lensch, Jim Gertsma, Leland Camp, and several others mapped in multiple Iowa counties as late as 2008. Those still living can attest to the physical toll of the work even though this never seemed to lessen their commitment.

Until 2004, the county soil surveys were printed in hard copy. They included the maps, official profile descriptions for each soil, tables of physical properties (texture, density, etc.), and interpretive tables indicating each soil's suitabilities and limitations for various uses such as sanitary facilities or building sites. In 2005, soil maps of the entire United States and all the accompanying text and tabular information became available online at WebSoilSurvey.gov, the official source for the most current information. At the website, a specific location can be searched in a variety of ways—by township, range and section, street address, longitude and latitude, and others. For several years, soil scientists led by Ryan Dermody in Waverly and Dan Pulido in Atlantic have been working full-time refining and updating portions of Iowa's soil maps for the National Cooperative Soil Survey, which maintains Web Soil Survey.

A soil map displays polygons showing the location and extent of different soil series and their phases in a given area, identified by symbols on the map (fig. 24). Phases are subdivisions of soil series that are important for the use and management of the soil. The most common two phases are based on the slope of the land surface—for example, 2 to 5 percent for gentle slopes or 14 to 18 percent for steep slopes—and the degree of erosion of the topsoil. Soil maps also show features that may be present such as rock outcrops, gravel pits, sandy spots, and sinkholes.

Across the state, hundreds of government, commercial, and nonprofit enterprises as well as private landowners rely on the maps, data, tables, and interpretations in Web Soil Survey every day to make planning decisions. It is important for anyone purchasing property in Iowa to consult the website for general information about the soils on a given property that could pose problems for farming, building foundations, basements, septic drain fields, landscaping, and other projects. It is important to recognize that the maps were not created at a fine enough resolution to always be accurate for every small corner of the land. It was of course impossible for soil mappers to probe every point on the ground; the smallest soil map unit on existing maps is about 2 acres. Instead, they drew lines onto aerial photos based on their conceptual models of the relationships between landforms and the soil types they had identified on the ground through probing.

Today's digital technology, including global positioning systems and geographic information systems, is making maps with much greater detail possible. For example, Bradley Miller of ISU's Department of Agronomy is leading a group currently working to improve the precision and accuracy of Iowa's soil maps. The group is able to delineate smaller landscape components than was previously possible by using the detailed LiDAR elevation data available for the entire state. LiDAR measurements—obtained by small aircraft that measure distance to the earth using a near-infrared laser—are accurate within 2 feet of

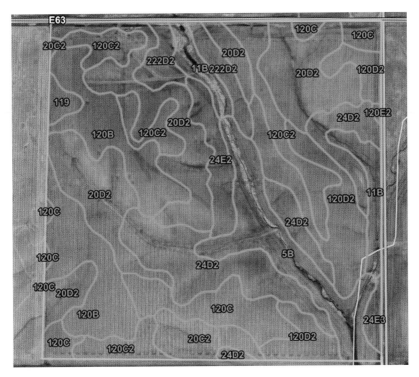

24. A soil map of one square mile of Marshall County, which features three soil orders and seven soil series. Letters signify slope phases—from B for gentle slopes up to E for very steep slopes. Numbers following letters indicate erosion phases—2 for moderately eroded and 3 for severely eroded. From the Web Soil Survey, USDA–Natural Resources Conservation Service.

elevation. In contrast, most older topographic maps of Iowa display contour intervals of 10 feet, which do not capture subtle features. By creating and refining a complex algorithm that combines LiDAR data with a large database of information about vegetation from remote sensing by satellites, the research group is better able to identify much smaller subfield areas with similar soil properties. According to Miller, this approach works because vegetation quality is a reflection of soil conditions—a concept recognized since ancient times.

Over many decades and continuing into the digital future, the work of hundreds of soil scientists, cartographers, technicians, and geographic information specialists has resulted in an enormous database of information about Iowa's soils—their classification, geographic distribution, properties, and potential uses and limitations. Now that you've seen how soil maps demonstrate the wonderful diversity of soils on our landscapes, the next chapter examines the role of the dynamic agents that have produced and continue to produce this rich tapestry.

CHAPTER THREE

The Stories They Can Tell

When the soil has been questioned, it will answer.
—ABBÉ JEAN COCHET, 1866

THERE IS SOMETHING EXCITING and deeply meaningful about peering into the body of the earth to see what lies hidden below the surface. As a geologist, I have always marveled at the unique awareness of time and the sense of connection with the past that come when we uncover the layers that make up the physical foundation on which we live out our lives. In a way, it reminds me of my time in Utah in the 1990s, when I had the opportunity to delve into the genealogy of my family. Poring over rolls of microfilm at Salt Lake City's Family History Library, I was excited to see our uncommon surname suddenly appear in handwritten cursive on a ship's passenger list—along with my great-grandfather's age, occupation, and hometown in Poland—followed by later discoveries about his parents and grandparents. What a treasure trove of information, generation passing to generation, rich with personal meaning.

In a similar way, witnessing the ground beneath my feet as it is revealed, layer by layer, always leaves me feeling a personal bond with the land and its history. Over the past twenty years, as I examined Iowa soils and glacial materials from more than 2,500 drill holes and backhoe pits, I rarely lost that sense of discovery and fascination. When I tired toward the end of a long day working in the elements, I just reminded myself that every time we drilled a hole or excavated a pit,

I was seeing something that humans had never before laid eyes on. Fortunately for our long-suffering drill rig operator, we were limited to a depth of 90 feet by the total length of the auger we had. When it could take the better part of a day to drill a 90-foot hole in 5-foot increments, it was tempting to want to go "just 5 more feet" since I'd never have another chance to see what was down there!

Our work didn't require us to drill very far into bedrock—just the top few feet of rock that was very weathered and fairly easy to penetrate without a rock-coring drill bit. We were literally only scratching the surface, of course, because the geologic strata that make up Iowa's bedrock are hundreds of feet thick. But while rock layers record the history of rock formation and the evolution of life over millions of years, soils exist on a different temporal and spatial scale. Compared to bedrock, Iowa's soil layer—the skin of the earth at the very top of the geologic column—is very thin and usually represents only several hundred to several thousand years of Earth's history. Because of its relative youth, however, it retains many more details about the natural history of those few millennia, whereas much detail about the past has been lost in the rock record as a result of millions of years of alteration from compaction, thorough decay of organic matter, changing groundwater chemistry, and other modifications. On the other hand, a surface soil is packed with complex and often subtle information that tells a story of how inanimate mineral matter and life have interacted since the soil began to form.

But soils at the surface of the earth have very different biographies depending on location. Where each soil is mapped, it is an individual body of nature with its own character and life history (Jenny 1984). In some unglaciated parts of the United States, such as the Southeast, the same soils have been at the surface for hundreds of thousands of years and belong to the order Ultisols, from the Latin *ultimus*, meaning "final." They are acidic, intensely weathered soils of low fertility in a very advanced stage of development. Compared to these geriatric soils, the oldest soils at the surface in Iowa are mere youngsters. In most of Iowa, they contain clues to the paleoenvironmental conditions that prevailed during the last 20,000 years at most. Soils in north-central

Iowa are even younger, because they began forming no more recently than about 11,000 years ago, after the last glacier had left Iowa for good. Most of them are even less mature than their age would suggest, for reasons I explain later in this chapter.

Deciphering the history of even such relatively young soils is not a straightforward proposition, however, because a soil at the surface is never a static end product. Soil formation continues as long as a soil isn't buried by geologic deposits or those related to human activities that are thick enough to halt processes like worm activity or root penetration. As a result, a surface soil is a record of change over the entire period of time it has lain at the land surface, with changes continuing into the present day. Given the radical transformations that Euro-American settlement and modern methods of tillage brought to the landscape in the past 150 years, these comparatively recent changes to the soil can be significant. That is why soil scientists have been documenting what they call dynamic soil properties, which are those characteristics of a soil that change with land use, management, and disturbance over the human timescale of decades to centuries. These include such measurable properties as the amount of organic carbon, the stability of the peds to resist degradation by rain and other forces, bulk density, and the pH of the topsoil, which is its degree of acidity or alkalinity.

Despite these dynamic properties, which become less pronounced with depth in the profile, soils have an inherent and intriguing ability to memorize the processes of their earlier stages of development. The Russians call this soil memory, with the memory bank consisting of a soil's morphology, meaning its horizons and more stable physical properties (Targulian and Goryachkin 2004). In this way, soils are an information storage medium with a truly long life span!

This chapter explains what and how Iowa's Mollisols, Alfisols, Inceptisols, Entisols, Histosols, and Vertisols "remember" about their past at a given point on the landscape. The five soil-forming factors introduced in chapter 2—parent material, climate, organisms, topography, and time—shaped these memories. The time factor is really more of an intangible continuum along which the other factors and soil-forming processes operate over the life of a soil, similar to how our own genetics and environments influence us throughout our own lifetimes. As more time passes, soils mature in predictable ways just as we do.

Parent Material, the Geologic Forebear of Soils

Parent material is perhaps the most fundamental of the soil-forming factors because it is the main control on a soil's texture. Texture in turn influences just about every aspect of soil formation. To understand what is meant by parent material, it is necessary to understand what Iowa's geologists have untangled over many decades about the sediments in which our state's soils formed. Many dedicated bedrock geologists have labored since the early nineteenth century to puzzle out the deep geology of Iowa, which of course is very interesting in its own right. From drilling and geophysical research, we know that the state is underlain at a great depth—1,000 to 4,000 feet below the surface except in the far northwest corner of Lyon County—by Precambrian igneous and metamorphic rocks billions of years old, nearly as old as Earth itself; that it was later submerged below shallow seas in which marine limestone, dolomite, and shale layers formed; and that at other times it lay exposed to the atmosphere, giving rise to sandstones, terrestrial shales, and mudstones. In his book *Iowa's Geological Past: Three Billion Years of Earth History*, Wayne Anderson presents the conclusions from two centuries of research to lay readers in a very accessible fashion.

Sedimentary rocks of Paleozoic and Mesozoic age are the parent material for some of the state's soils, especially in northeast and parts of southeast Iowa. But the majority of the state's soils developed in glacially derived sediments laid down by the thick continental ice sheets that overran Iowa at least seven times between 2.7 million and 500,000 years ago (dates codified in Cohen and Gibbard 2019) called the Pre-Illinoian tills; by Illinoian and Wisconsinan glaciers that together overran portions of the state at least four times in the past 300,000 years (Tassier-Surine et al. 2018; Kerr et al. 2019); and by strong

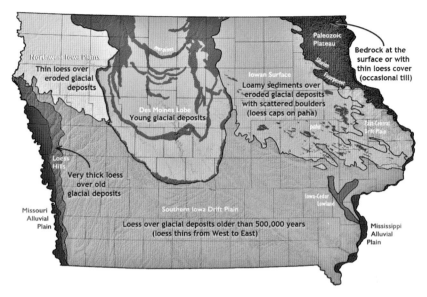

25. Landform regions of Iowa. After Jean Prior, *Landforms of Iowa*. Courtesy of the Iowa Geological Survey.

winds, which deposited silt and clay plucked from floodplains that had amassed large volumes of fine-grained sediments delivered there by glacial meltwater. Other soils formed in the sediments laid down by rivers and smaller streams. Jean Prior's *Landforms of Iowa* describes in splendid detail the nature and origins of the landform regions of our state. They range from the oldest, the Paleozoic Plateau in northeast Iowa, to the youngest, the Mississippi and Missouri alluvial plains. The map of the landform regions in figure 25 also indicates the main soil parent materials across the state.

Deposition of the thick uppermost layer of loess that blankets much of Iowa, called the Peoria Formation, began in earnest about 27,000 years ago during the Late Wisconsinan glacial stage. Between 45,000 and 16,500 years ago, the continent-scale Laurentide ice sheet covered nearly all of Canada and advanced southward into the present United States at least twice. It reached its maximum southern extent about 18,000 years ago, then gradually retreated, but two smaller ice lobes—the Des Moines Lobe in north-central Iowa and the James Lobe in North Dakota—again surged ahead from the ice front at least three times between 16,500 and 14,000 years ago.

While many places in the U.S. brag about having the oldest something, Iowans on the Des Moines Lobe have the distinction of living on one of the youngest and freshest landscapes in the country. At the same time, they also have something very old, because the young glacial till of the Des Moines Lobe contains cobbles and boulders of some of the oldest rocks on Earth, brought here from the Canadian Shield, the original nucleus of the North American continent. The marvelous shaded relief map of the midsection of the North American continent in figure 26, compiled from more than 20,000 digital elevation maps at the University of Minnesota, shows what this part of the world would look like from space with its vegetation and cultural features removed. The very young deposits and landforms of the Des Moines Lobe and the James Lobe—along with the clear path they took from the north-northwest—appear smooth because they are much less dissected by streams than are older parts of the landscape.

Loess of the Peoria Formation, a widespread side effect of Late Wisconsinan glaciation, was deposited before the Des Moines Lobe advanced. Loess now blankets most of the state except for the Des Moines Lobe—where the glacier stripped it off or debris melting out of the ice buried it—and large parts of the northeast quarter of Iowa called the Iowan Erosion Surface. On the Erosion Surface, erosion had already removed the loess by about 16,000 years ago during a period of maximum cold in northern Iowa. During that time, freeze-thaw of the unglaciated but deeply frozen ground—or permafrost—loosened the surface deposits, and gravity moved several feet of the saturated materials downslope over time.

In general, the loess becomes thinner in an eastward direction and varies from nearly 200 feet thick in parts of the Loess Hills of western Iowa to less than a few feet thick in parts of eastern Iowa. Studies show that the rate of loess deposition over thousands of years probably averaged only 1 to 2 millimeters per year in western Iowa and even less farther east, which nevertheless translates to more than 75 feet over

26. Topographic map of the upper midwestern United States and part of Canada. Courtesy of Paul Morin, Polar Geospatial Center, University of Minnesota. The image has been cropped to show detail; outline of Iowa by the author.

15,000 years! But it did not fall at a steady rate (Bettis et al. 2003). At times this dust may have accumulated in western Iowa as rapidly as 6 inches in 10 years, which was too fast for soil-forming processes to keep up. That may not sound like a lot unless it's collecting on your furniture, but it's enough to choke off and bury a soil. However, whenever the dust fall stopped or slowed significantly for a long enough period of time, soil formation was able to proceed. The A horizon simply built upward and the new material was incorporated through the activities of soil fauna, along with alternating freezing and thawing and shrinking and swelling.

Iowa's hundreds of soil series can be grouped according to their landform region, parent materials (type, age, and thickness), topographic features, and other characteristics (see fig. 25). The online "Highway

Guide of Iowa Soil Associations" groups them into twenty-two soil associations, most of which have some degree of loess cover ranging from a few inches to a few hundred feet (Natural Resources Conservation Service 2008).

Climate, Vegetation, and Fauna through Time

Unlike the majority of Iowa's soil parent materials, which have remained relatively constant for at least the past 10,000 years or so, the vegetation that took hold in these young soils has fluctuated notably with climatic changes. Climate—the long-term temperature patterns and amounts and timing of precipitation—is probably the most dynamic, that is, the most changeable factor of soil formation. If we disregard short-term changes on the scale of a few years or decades, climate is the main influence on the types of vegetation and fauna that thrive at any given location. Fluctuating long-term climatic patterns in the midcontinent region over the last 16,000 years or so caused plant and animal communities to migrate back and forth in a general north-south direction.

The fossil record, consisting of plant pollen and seeds along with beetle, snail, and small rodent remains, shows that during the time of maximum cold between about 25,000 and 16,500 years ago, spruce forests followed by arctic tundra dominated the landscape (Prior et al. 1982; Mutel 2008). Tundra, a treeless ecosystem in which the subsoil remains frozen throughout the year, may have held on even later on the Des Moines Lobe. Many of the mammals that roamed the land during Iowa's last ice age were like nothing alive today. Since the early 2000s, there have been at least three exciting discoveries of bones from extinct ice age mammals in Iowa—giant ground sloths near Shenandoah, mammoths near Oskaloosa, and a giant short-faced bear near Atlantic, all of which were much larger than their modern counterparts. You can discover for yourself just how big they were at the University of Iowa's Museum of Natural History.

But as the Des Moines Lobe began to retreat, the climate moderated and the megamammals disappeared. Boreal forests of spruce, larch, and pine became established, followed by deciduous forests of oak, elm, and ash by at least 12,000 years ago (Eilers and Roosa 1994). By 9,000 years ago, the climate had become much warmer and drier and prairie grasslands had spread from west to east, reaching eastern Iowa roughly 5,500 years ago. About 4,200 years ago, Iowa and most of the U.S. Midwest experienced decades-long drought conditions (Booth et al. 2005). Abundant charcoal from bogs and fens, dated to between 6,200 and 5,000 years ago, indicates that many depressions on the Des Moines Lobe dried up at times, and widespread fires were frequent (Baker et al. 1992). The rain that did fall was intense and erosive and most likely set back soil formation on the denuded landscape.

Since about 4,000 years ago, the climate in eastern and central Iowa has alternated between moist and dry conditions and has nurtured a shifting prairie-woodland mosaic. In parts of the state, this probably took the form at times of savanna, a mix of prairie and oak (Baker et al. 1992). In areas that remained under deciduous forest or woodland for the majority of that time, Alfisols like the Fayette soils and some seventy-five other series formed on the landscape. But the majority of the state as recently as 200 years ago was covered by prairie and probably resembled this reflection by Iowa's well-known nature writer Larry Stone: "I visualized a sea of head-high stems, waving across the rolling plain. Bison grazed in small herds on the horizon and prairie chickens loafed on the knolls. The sky swarmed with waterfowl above a pothole left from a glacier's retreat 10,000 years before. The expanse was dotted with color from yellow sunflowers, purple blazing stars and green rattlesnake masters poking their heads between the clumps of prairie grass" (1999).

In this history of continual environmental change, the most rapid and dynamic transformation of Iowa's landscape has occurred in the last 150 years, as Euro-American settlement and widespread intensive agriculture have taken hold. In that short time, the dominance of nonnative species of crops and weeds has had a profound impact on the soils of Iowa.

The Importance of Topography

In addition to parent material, climate, and vegetation, another major influence on soil formation is the topographic position of a soil on the landscape. Recall Robert Ruhe's slope model of summit, shoulder, backslope, footslope, and toeslope from figure 9 in chapter 1. When climate and robust vegetation are relatively constant over long periods of time, most of these landscape positions can also be fairly stable. Erosion and deposition by glacial ice are absent, and erosion and deposition by water and wind are not extreme or widespread. One exception is the toeslope, which stream deposits can bury repeatedly even during short-term precipitation events. If the deposit is thick enough, soil formation has to start from the top all over again, which helps explain the abundance of young soils—Entisols like the Nodaway series highlighted in chapter 2—on floodplains across much of the state.

However, anything that brings about a major change, particularly a change in vegetation density or root depth, can trigger widespread erosion. Erosion may strip off the topsoil or even most of the soil profile, while deposition can bury a soil and stop the process of soil formation. For example, in the past 4,000 years, many small and moderate-sized valleys in western Iowa underwent at least four major cycles and numerous minor cycles of gully erosion alternating with sediment infilling, called cut-and-fill cycles (Bettis and Thompson 1981). Downcutting in a valley often works its way headward up the tributary streams, which in turn initiates erosion on the slopes within the watershed.

The latest major cycle of downcutting in western Iowa began in the mid-1800s with Euro-American settlement and the intensification of agriculture, which severely affected the entire state. As the settlers converted millions of acres from prairie to row crop or woodland to pastureland, erosion on the shoulder and backslope positions became prevalent. One result is that many eroded loess slopes in western Iowa now feature Inceptisols or even Entisols like the Ida soils, whose very young profiles consist of an Ap horizon just an inch or two thick resting directly on top of the C horizon (fig. 27). (Compare this Ida profile to the Tama profile in figure 2 in chapter 1, which is probably what Ida

27. Ida soil profile, an Entisol from Monona County. Courtesy of the USDA–Natural Resources Conservation Service.

soils looked like 150 years ago.) Eroded topsoil from the shoulder or backslope ends up on the footslope and toeslope positions, and some of it gets carried farther in the next few storms (fig. 28). Too much of this sediment finds its way into Iowa's rivers and lakes, a process that continues unabated today, and some is eventually carried to the Mississippi delta, as we'll see in chapter 6.

As you have probably realized by now, the soil-forming factors clearly are not independent of one another. Wherever all the soil-forming factors are the same, the soils will be very similar, but if one of the factors is sufficiently different, the soils will differ. For instance, soils forming

28. On this farmland in western Iowa, the light-colored slopes are highly eroded, while the dark bottomlands are sites of deposition. Photograph by Tim McCabe, USDA–Natural Resources Conservation Service.

in identical parent material under similar vegetation during periods of landscape stability but on different slope positions can differ considerably in the thickness of horizons and in drainage characteristics and color. Just as important as slope position are slope steepness, length and curvature of the slope, and slope aspect—the compass direction it faces. Almost all soil series in U.S. Department of Agriculture soil surveys are further subdivided into soil phases.

The most common breakdown into soil phases is based on slope steepness, which has a significant influence on soil properties and considerable impacts on agricultural practices and other uses. In most Iowa counties, the USDA slope categories are 0 to 2, 2 to 5, 5 to 9, 9 to 14, 14 to 18, and 18 to 25 percent, where a 5 percent slope, for example, means that the ground surface drops 5 feet over a horizontal distance of 100 feet. Slope steepness is important first and foremost because it is the main thing determining whether surface water from rain or snowmelt infiltrates into the ground or runs down the slope, potentially causing erosion. Better infiltration of water also results in more strongly developed soils, for example, soils with thicker horizons and better-developed structure. In general, steeper slopes tend to have thinner and less developed soils.

Soil Processes through Time

Equipped with this basic understanding of the environmental conditions under which a soil develops, let's now look more closely at the pedogenic processes that occur over time within a soil as it ages. The longer a soil resides at the surface of the earth, the more mature it becomes and, in general, the deeper it extends. Starting from bedrock, it takes at least 10,000 years for a soil to develop a B horizon with good blocky structure and an accumulation of clays and iron oxides. Of course, most of Iowa's Alfisols and Mollisols formed in unconsolidated loess and glacial deposits rather than in bedrock. From paleoenvironmental studies, we can surmise that they began forming between 11,000 and 5,500 years ago, depending on location. They continue to form today by adjusting to current conditions.

One of the four fundamental processes that occurs during soil formation is the addition of organic material to the mineral substance of the parent material, whether that parent material is weathered rock, glacial till, loess, or stream deposits. It is the dominant and defining soil-forming process in the case of prairie soils, where long-living deep roots and thick rich topsoil sustain one another like milkweed and monarch butterflies. Incredibly, the roots of perennial prairie plants can extend 15 or more feet into the ground.

Rooting depth varies a lot among plant species, and the prairie plant community differs according to the soil and its position on the landscape. Several years ago my partner and I owned 10 acres in Jasper County, including 5 acres of hilly pasture that we seeded with a native prairie mix. We were able to observe firsthand how different species thrived on different soil types, from silt loam soils in loess to clay loam and loamy sand soils in till. Both the parent material and the slope position influenced the moisture content of the soil and consequently the plant community: from drought-tolerant plants like round-headed bush clover in the dry sandy soils to mesic plants in the loamy soils and water-loving sedges in the more clay-rich soils near the bottom of slopes.

Prairie roots add organic matter to the soil every year. Even though the majority of perennial roots remain living throughout the year, some of these fibrous roots and most of the root hairs die and decay each winter. Along with the residues of plant parts that annually fall to the ground, the roots are broken down by soil microbes, soil insects, earthworms, and rodents that live in the soil. Night crawlers are very efficient at retrieving plant material from the surface, then digesting the organic matter along with minerals to form simpler compounds. Centuries ago, Aristotle recognized the importance of earthworms, and whether he called them "the intestines of the earth" as is often quoted, the analogy is certainly apt. Earthworms along with other fauna like ants, potworms, springtails, and fly larvae are responsible for the beautiful granular soil structure we see throughout the entire A horizon of many Mollisols like the Tama soils, which allows a clod to crumble easily in the hand, similar to dry cottage cheese. Over hundreds to thousands of years, the addition of plant matter to the root zone enriched the upper part of Iowa's Mollisols. At times, dust or soil eroded off the slopes above was slowly added to the topsoil by wind, water, or gravity, causing the A horizon to slowly thicken as soil formation kept pace.

Unlike with the black Mollisols, the addition of organic matter plays much less of a role in the formation of Alfisols. Recall the Fayette soil profile from chapter 2, with its grayish brown A horizon no more than 4 inches thick. Although the addition of organic matter from aboveground each year isn't all that different between prairie and woodland soils, the portion produced underground is much less under trees. So while Mollisols are defined by their thick A horizon, a soil is considered an Alfisol because of its B horizon, specifically the Bt horizon that I introduced in chapter 2. This clay-enriched horizon forms as a result of another of the four processes that transform parent material into soil, namely, the translocation of material from one part of the profile to another.

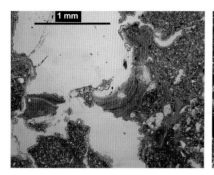 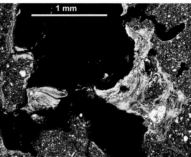

29. Laminated clay coating on the surface of a large soil pore. Photographs by the author.

Bt horizons are present in some Mollisol profiles as well, specifically Argiudolls such as the Tama soils. In both Alfisols and Mollisols, clay translocated to the Bt horizon is concentrated in laminated coatings on the surfaces of blocky peds and along worm passages, root channels, and other soil pores. These coatings can usually be seen with a 5X or 10X hand lens as glossy gray surfaces, but the microlaminations are visible only with a microscope. The thick clay coating in figure 29 formed on the wall of a large pore and is made up of about a dozen individual layers—most apparent in the left view—each of which represents a separate episode of clay translocation. It is the parallel orientation of the plate-like clay particles in the coating that causes the bright yellow and orange colors under circularly polarized light seen on the right.

But before clay can move down through the soil profile, calcium carbonate and other dissolved minerals must be leached from the A and B horizons and carried down into the C horizon. When they were deposited, most parent materials in Iowa—including loess and glacial till—contained abundant tiny calcite particles in their matrix of silt and clay in addition to sand-sized or larger grains of calcite in the till. The ice sheets that marched into Iowa from the north and northwest had plucked these minerals from limestone, dolomite, and shale bedrock as well as the older till layers they had overrun. Once the carbonates are dissolved and leached from a soil horizon, the soil matrix no longer effervesces when treated with a drop of 10 percent hydrochloric acid, a small bottle of which is a necessary tool for evaluating soils in the field.

This loss of material from one or more soil horizons is another major process of soil formation. Erosion removes the finer particles in the topsoil, and some organic matter is lost to decomposition on a regular basis. Leaching of carbonates by percolating water lowers the pH of the soil, which is significant because pH controls most of the physical and chemical processes of soil formation. Values for soil pH in Iowa's cultivated soils, before lime or sulfur amendments, range from 4.5 (very acidic) to 8.2 (very alkaline), with 7 being neutral (Hannan 2017). Corn and soybean crops do best with a pH of at least 6.5, but the application of chemical fertilizers like anhydrous ammonia and other forms of nitrogen, along with the removal of crop residues, causes soil pH values to decrease to levels below 6. At that point, many farmers add ag lime—calcium carbonate—to the soil to bring the pH back up (Mallarino, Sawyer, and Barnhart 2013).

The buildup of clays in a soil's B horizon begins along shrinkage cracks and channels made by roots and soil insects and earthworms. As the soil dries out and is rewetted during successive seasons, expandable clays cause it to alternately shrink and swell. Expandable clays include a group of clay minerals called smectite, which is very common in Iowa's glacial and windblown deposits. Shrinking and swelling of these clays give B horizons their typical blocky soil structure, and these blocky peds provide convenient deposition sites for even more clay. Along with translocation of clay, brand-new clays can form within the B horizon from the fourth major soil-forming process— the weathering and transformation of other minerals such as mica and feldspar.

In figure 29 (right side), orange colors in the coating indicate that at times iron moved down along with the clay. This movement of iron in solution is another example of translocation during soil formation, which can radically alter the appearance of a soil profile. Iron stains the clay coatings on ped surfaces and may also precipitate as orange concentrations of iron oxides. These may be uncemented soft masses or concretions or nodules of varying degrees of hardness. (A concretion consists of concentric layers like an onion, while a nodule is amorphous, more like a potato.) Soil scientists refer to these concentrations as redox features, a concatenation of "reduction" and "oxidation." This implies the chemical change of iron or manganese minerals through the biochemical process of reduction during water saturation or oxidation from exposure to air. Redox features are also commonly called mottles.

Redox features occur only at depths in the soil that experience saturation for a sufficient period of time, leading to reduction, followed by drying out of the soil, leading to oxidation. In plain terms, the zone in the soil profile that displays redox features is the zone where the water table fluctuates or has fluctuated in the past. So, for example, if iron oxides are common within the upper 2 feet of the ground surface, this implies that the soil is at times saturated at that depth for more than a few days. On the other hand, if such redox features are absent or rare in the upper few feet, we know that the soil is saturated only for very short periods of time after rain events and the water table never rises very high in the profile. In these well-drained soils, such as a Tama or a Fayette soil, water moves through the profile readily—slowly enough that it is available to plants for most of the growing season but quickly enough that saturation doesn't inhibit the growth of roots. These are ideal soils for growing crops and gardens and have relatively few limitations for other uses, such as basements, where a shallow water table can create serious seepage problems.

Many soils with good drainage nevertheless may have another type of redox feature fairly high in the profile, namely, tiny black manganese oxide concretions or thin coatings. Because manganese reduces more readily than iron, the soil has to be saturated only for a few days following rainfall for manganese oxide concretions to form when it dries out sufficiently. If iron oxides are absent, manganese oxides alone do not imply a fluctuating water table.

In the lower B and C horizons, many Iowa soils exhibit features with yet another color related to the behavior of groundwater in the profile. These redox depletions are spots or stains indicating periods of saturation lasting long enough to have led to the complete removal of iron and thus to decreased pigmentation in the soil. They have grayish colors such as 10YR 5/1 or 5/2, called low-chroma colors. Generally, the more of these gray depletion features that are present and the larger they are, the more frequent or longer were the periods of saturation. This can be an important piece of information in our state, where manure storage tanks and ponds are becoming more and more commonplace on the landscape. These are just two examples of land uses that are appropriate only on or in soils with a relatively deep water table to ensure against groundwater contamination. The relatively shallow presence of redox depletions in a soil is a clear warning sign that the water table rises high into the profile during certain times of the year.

A good example of a soil with colors that indicate a shallow water table is the Nicollet soil series. There are nearly 1.1 million acres of Nicollet soils on level till plains in every county of the Des Moines Lobe, making it the fifth most abundant soil series in the state. They formed in loamy glacial tills of the Dows Formation deposited during the Wisconsinan glacial stage—parent materials that are geologically speaking very young, between 12,000 and 14,000 years old from north to south.

The thick topsoil of a Nicollet soil indicates its formation under tallgrass prairie, although today these soils are nearly all planted to corn and soybeans. Still, their black A horizons can be impressively thick, ranging from 10 to 24 inches (fig. 30). Nicollet soils do not have Bt horizons. Instead, immediately below the A horizon there is a Bw horizon, in which the *w* signifies that it is weakly developed but different in color or structure from the overlying horizon. Below the Bw is a Bg horizon, where the *g* signifies a gleyed matrix with gray, bluish gray, or greenish gray colors. (The word "gley" comes from a Ukrainian or Russian word for a sticky blue clay.) This particular Bg horizon has a color that corresponds to the 2.5Y page on the Munsell charts instead of the 10YR page like the overlying horizons. If you were to look at the Munsell charts for 2.5Y and 10YR, you might not notice much of a difference between them, although it is easier to see over a larger surface in the field. Nevertheless, this subtle color difference carries meaning. It tells us that in the past, the water table at that location frequently

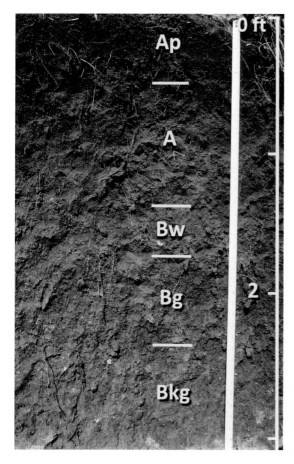

30. Nicollet soil profile, a Mollisol from Boone County. Photograph by Gerald Miller.

rose to the top of the Bg horizon, within less than 20 inches of the surface, for significant lengths of time. As a result, much of the iron in the fine-grained portion of the mineral matrix has been reduced and lost by leaching.

Although not visible in figure 30, Nicollet soils also usually contain a few small olive brown iron oxide redox features in the Bg horizon, from which we're able to deduce that the water table does drop at times, probably toward the end of the growing season. Nicollet soils do not typically exhibit black manganese concretions, which require a longer dry period for oxidation than does iron, so we can surmise that the water table doesn't stay down for very long.

The soil in the photo also contains abundant carbonate concretions and nodules—tiny white particles—at a fairly shallow depth in the horizon labeled Bkg. The calcium carbonate was leached out of the matrix of the overlying horizons and then precipitated from this solution as secondary or pedogenic carbonates. Generally, the deeper such carbonates occur in the profile, the deeper the prevailing water table is situated. In summary, to a knowledgeable observer the colors, oxides, and carbonates are all clues to the pattern of water saturation in the profile.

The explanation for why Nicollet and related soils have a high water table is mainly geologic in nature. Because the landscape of the Des Moines Lobe is so young, geologically speaking, there hasn't been sufficient time for a mature network of river valleys and tributary channels to develop and lower the water table, as there has been on Iowa's older landscapes. As a result, the water table on the Lobe is typically shallow, and many of its soils have gleyed C horizons. These include poorly drained Aquolls like the Webster and Okoboji series as well as the Histosols. Furthermore, both flat topography and a shallow water table reduce water movement, which in turn inhibits leaching of carbonates and therefore the translocation of clay. This may partly explain the youthful appearance of many soils on the Des Moines Lobe, very few of which have a Bt horizon.

The Histosols and some of the Aquolls on the Lobe are found in the closed depressions of prairie potholes. Potholes are smaller and much shallower than glacial kettles like Storm Lake and Spirit Lake, which are deep, steep-sided depressions that formed when an ice block buried by glacial deposits melted and left a pit on the landscape. Most potholes probably formed when the stagnating Des Moines Lobe ice sheet, riddled with cavities and tunnels, slowly melted in place and dropped its debris to form an irregular surface with thousands of depressions an acre or smaller in size. They are like shallow bowls, from which water cannot escape overland but must percolate down into the soil profile. One type of wetland, they were once one of the most diverse ecosystems in the Upper Midwest.

Most of Iowa's potholes and a few kettles have been drained and are now farmed. Where we find the water table deeper than a foot or two below the land surface, this is most likely due to artificial drainage for the purposes of agriculture. More than 30 percent of Iowa cropland is now drained with underground perforated pipes. These drainage tiles are mostly concentrated on the Des Moines Lobe, but across the entire state there are an estimated 2 million miles of them (Takle and Gutowski 2020). One study found 59 miles of tile in just 2 square miles of Hamilton County (Hoyer 2011). Despite this extensive artificial drainage, the water table is still relatively shallow in most places. Standing in a field in north-central Iowa after a rainstorm, surrounded by shallow depressions filled with water before it drained into the tile intakes, I sometimes imagined the presence of a quiet ocean of groundwater beneath my feet. Of course it is there below Earth's land surface everywhere, unseen but constituting 30 percent of the planet's fresh water, second only to glaciers. However, it lingers especially close on the Des Moines Lobe and other recently glaciated parts of the Upper Midwest.

You may not be able to see these farmed potholes from your car window when the crops are high, but their presence is still very apparent on soil maps, where they show up as thousands of small circular or lobed polygons mapped as hydric soils. These soils retain the memory of their former soggy state in their physical properties. Identifying hydric soils is critical to delineating and restoring former wetlands in Iowa, so they are of great interest to soil scientists and many other environmental scientists in a state where wetlands are an essential ecosystem.

Soil memory is a complex concept, because a soil's morphology can change in at least two primary ways. One way is simply by aging, because even a soil residing at a stable land surface with stable vegetation generally becomes thicker and develops more horizons with the passing of time, since more additions, losses, translocations, and transformations take place in the profile as it matures. Another way is in response to major changes in climate or vegetation. In this latter case, the soil profile can undergo chemical changes as well as physical changes in the makeup or thickness of its horizons as well as changes to its structure, colors, coatings, and even texture in some cases. Iowa's soils are still adjusting to the most recent change in vegetation, the rapid transformation of Iowa's prairie to farmland between 1850 and 1900. What really makes soil sleuthing challenging (and fun) is the fact that much of the new information is superimposed over what is left of the old. Some have described such a soil as a palimpsest, a multi-layered record similar to a parchment or canvas whose original text or art has been partially erased and then overwritten.

The older a soil becomes, the more environmental changes it experiences. Since most of Iowa's parent materials are at least several thousand years old, the soils formed in them have undergone many changes in climate and vegetation, as outlined at the beginning of this chapter. So perhaps they could all be said to be palimpsests, although in many cases the earlier traces are very subtle or completely erased. However, one clear example is the Downs series, an eastern Iowa Alfisol with a silty E horizon. Even though it is generally found today under grasslands or crops, it is in a part of the state that has seen shifting prairie-woodland vegetation for the past 4,000 years. Downs soils are considered transitional soils between the Tama Mollisols and the Fayette Alfisols, all three of which formed in loess.

At 7 to 8 inches thick, the A horizon of a Downs soil—now an Ap horizon after decades of cultivation—is much thicker than that of typical Alfisols but not thick enough for the soil to qualify as a Mollisol. So it is classified as a Mollic Hapludalf (fig. 31). It is very possible that the E horizon and strongly developed Bt horizon were developing under forest or savanna vegetation as early as 10,000 years ago, before the prairie reached eastern Iowa. If not for the presence of an E horizon, another interpretation might be that Downs soils are former Mollisols—Argiudolls—with a Bt horizon and an eroded A horizon.

Distinguishing physical features from different periods of soil formation benefits greatly from the use of soil micromorphology—viewing and describing thin sections of soil under a microscope. This is especially true of Iowa's paleosols. Some of these ancient soils may have lain at the land surface for longer than 100,000 years, during which time their features were overwritten repeatedly due to major

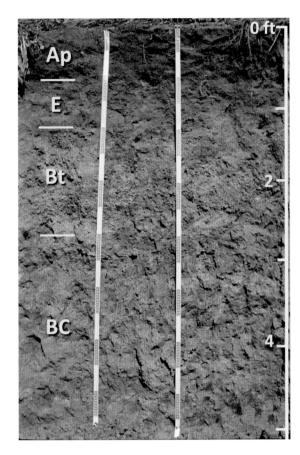

Ap

E

Bt

BC

0 ft

2

4

31. Downs soil pro-
file, an Alfisol from
Winneshiek County.
Photograph by Jon
Sandor.

environmental shifts. When younger glacial or loess deposits finally
buried them, soil formation ceased. However, over the past centuries
and mainly during the last 150 years, erosion has exposed them at the
surface once again in many places. As a result, in these areas paleosols
make up all or part of the parent material for the new surface soils,
whose features are imprinted over the old by modern processes. Soils
that are developing in these ancient soils are only the latest layer of
paint applied to Earth's canvas.

CHAPTER FOUR

Soils on Iowa's Hidden Landscapes

The truths of the earth continually wait.
—WALT WHITMAN, 1856

THE SOILS THAT GRACE today's landscapes reflect the parent materials and ecosystems that have existed throughout their time of formation, and it is the shallow surface geology of each of Iowa's landform regions that defines soil parent material. Even though most of our present-day soils exist today in an agricultural or urban ecosystem, nevertheless they contain clues to the environmental conditions that prevailed during the last several thousand years in this part of the world or during the last few hundred years in the case of soils in many of our stream valleys.

Of course, several thousand years comprise only the last few sentences in Iowa's natural biography—a geologic history that spans hundreds of millions of years. Regardless of which geologic material lies at the surface, it is only the very top layer of the geologic column, the icing on a massive layer cake composed of sedimentary rock strata and layers of older loess and glacial till. Beneath the icing, there are as many former ancient land surfaces as you have the knowledge and patience to figuratively dive into the cake and seek out. Every horizontal plane you can see in the rock column that was not submerged beneath shallow seas was at one time a subaerial surface exposed to the atmosphere—perhaps for a few days, perhaps for a few centuries or even millennia.

On some of these surfaces, soils formed to one degree or another before they were either buried by younger deposits or eroded away. Those land surfaces that escaped erosion, which lay exposed for thousands of years and were eventually buried, exhibit truly ancient paleosols. These are usually lithified, which means that over time they turned into rock, often mudstone. Others have studied and described paleosols in the bedrock column around the world and in the United States (Retallack 2019). Although such paleosols are certainly present in the bedrock of Iowa, most are buried at considerable depth by glacial deposits, and we see them only in drill cores. A few are exposed in limestone quarries, including paleosols in Pennsylvanian mudstones of Union and Decatur Counties, in Mississippian shales and claystones of Keokuk County, and a few others (Pope and Marshall 2010; Witzke et al. 1990).

But rather than focusing on paleosols preserved in rock, this chapter shines a light on buried soils found within the unlithified sediments resting on Iowa's bedrock. Before burial, these soils formed in layers of glacial till and silty loess from ice ages of the past 2.5 million years or so—the so-called Pleistocene Epoch of geologic time. Pleistocene paleosols are present in marvelous variety in Iowa, including some almost as dark in color as modern soils and one across most of southern Iowa that is more than 25 feet thick in places. They are our best and usually our only record of the interglacial stages—the times between ice ages when Iowa's landscapes experienced wetter, warmer, or drier climates than we are experiencing during our present interglacial times. For the most part, these paleosols and their landscapes are hidden from our view except where erosion has exhumed them, exposing small keyholes into the past.

There are much younger paleosols buried in many of Iowa's stream valleys that hold important clues to the environments and landscape processes of the past several thousand years, called the Holocene Epoch. Because their time at the land surface before burial overlapped the presence of humans in the Midwest, these soils are a signal to archaeologists to look for artifacts at those same levels. At sites where early Native Americans lived, worked, and played, they left behind evidence of their occupation to be preserved in the same soils that provided them a floor and the necessary resources for living.

Although some of the Pleistocene and Holocene paleosols of Iowa are very dense—that is, with high bulk density—they have not been cemented or compacted into sedimentary rocks like shale or mudstone. As a result, they fortunately exhibit many features that correspond to what we see in our modern-day soils. Thus, we are able to study old soils and reconstruct some of the environmental conditions under which they formed many millennia ago.

Something Old, Something Relatively New

Before examining what paleosols can tell us about the past, however, it's important to realize that they continue to affect us even today. Southern Iowa farmers who may be unfamiliar with the term "Pleistocene" are nevertheless very much acquainted with these stubborn old soils. They sometimes use the word "gumbo" to describe areas in their fields with dense clay soils that are typically wet and extremely sticky during planting season, often hard and dry during the summer, and much less productive than the rest of their acres. Often these are locations for seeps where groundwater oozes out, perfect traps for getting that tractor stuck (fig. 32).

On the long rolling hills that give southern Iowa its characteristic appearance of a Grant Wood painting, these horizontal bands of gumbo mark places where erosion removed all or almost all of the overlying loess or till layer and uncovered or exhumed these ancient soils. New soil properties were then superimposed onto the paleosols in response to the changed conditions of climate, vegetation, and topography. However, several important properties of these next-generation soils, particularly in their B horizons, were inherited from the older generation's parent soil—such things as texture, density, mineral composition, and even soil structure in many cases. Physical features shaped by modern environmental conditions—carbon content, color of the surface horizon, some chemical properties like pH, and others—became imprinted onto soil features from the past,

seep

32. Groundwater seeping out above a dense clay-rich paleosol in Adams County. Photograph by the author.

making for very complex soil profiles. Separating out and attributing the properties of these palimpsest profiles to influences from ancient times, the more recent past, or the present can be very challenging or admittedly impossible at times.

In a profile of a Clarinda soil from Adams County, the B horizons extend deeper than the 5 feet shown, in parent material comprised of a Pleistocene paleosol with the texture and consistence of stiff clay (fig. 33). (The initial "2" in the designation "2Bt" means that horizon's parent material is the second one encountered below the surface.) In contrast, the profile's A horizon was developed in more recent loess that has less than one-third the clay content of the paleosol. Although not always present, sediments overlying these paleosols are either loess or a mix of loess and glacial till material that was eroded higher upslope, transported, and redeposited.

This soil is classified as an Aquoll because of its poor drainage, indicated by the gleyed colors, and an Argiaquoll because of its thick clay-rich Bt horizons. Containing more than 50 percent clay, these horizons are nearly impermeable to water. So rainwater percolating

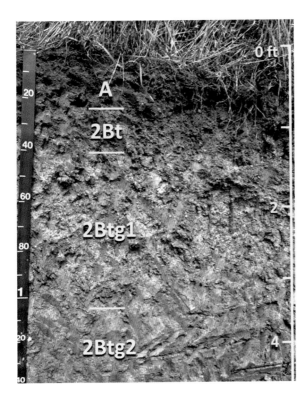

33. Clarinda soil profile, Adams County. Courtesy of the USDA–Natural Resources Conservation Service.

down through the upper horizon perches temporarily on the paleosol, then flows laterally and exits at hillside seeps like the one shown in figure 32.

In southern Iowa counties like Appanoose, Clarke, Lucas, and Wayne, these next-generation soils rooted in ancient paleosols cover more than a quarter of today's land surface. In much of Clarke County, surface soils that developed in paleosol parent material make up nearly 30 percent of the county's modern-day soils. In many other counties of the Southern Iowa Drift Plain—shown in figure 25 in chapter 3—they comprise 10 to 15 percent of the surface soils, with series names like Armstrong, Bucknell, Lamoni, and Clarinda.

The poor condition of the crops growing on these clay-rich soils is visible evidence of their low fertility and poor tilth. Driving by in May or June before the crop canopy grows tall, you can spot the sparse, stunted, and yellowed condition of corn or soybeans planted on these soils. The Corn Suitability Ratings for Clarinda, Armstrong, Bucknell, and Lamoni soils reflect this, as they range from 5 to 47 compared to ratings in the 70s and 80s for loess soils. In most cases, the only reason they're seeded at all is because avoiding them with the multiple-row planters in use today is more trouble than it's worth.

Another troublesome property of these clay-rich paleosols is their penchant for swelling as they take up water and shrinking when they dry out. This is due to the type of clay minerals they're so rich in, namely, smectite. The clays in these southern Iowa paleosols occur as microscopic aggregates, like very tiny books containing thousands of individual platy crystals. These platy clusters are negatively charged and swell by attracting and adding layers of water molecules between stacks of submicroscopic plates. The shrink-swell potential of each soil series is given on WebSoilSurvey.gov in the tab for soil physical prop- erties under something called a rating of linear extensibility. This is the volume change that a soil clod undergoes when it is dried in the laboratory. Clarinda soils might have a linear extensibility rating of 11, compared to only 3 or 4 for soils formed in loess. Expansive soils that shrink and swell to this degree can be a serious hazard to structures because rigid foundations are prone to cracking as the volume of the soil supporting them changes in response to moisture. Although much less dramatic than natural disasters such as hurricanes or floods, the damage that expansive soils cause to buildings, driveways, highways, walls, swimming pools, and other structures is one of the nation's most costly and ongoing natural hazards.

Despite the problematic nature of these Pleistocene paleosols for farmers, homeowners, and others, they are beneficial to us in other ways. In much of the state, clay-rich paleosols behave as aquicludes, protecting the groundwater in layers of more permeable soil and rock beneath them. An aquiclude is a layer of earth through which water moves extremely slowly or not at all. Because of their high clay content and swelling properties, the Pleistocene paleosols are excellent aqui- cludes as long as they don't dry out.

Archives of the Geologic Past

Geologists have learned much about Iowa's glacial and interglacial times by identifying and studying buried paleosols. Just as Iowa's uppermost geologic layers are the parent materials for our state's modern-day soils, deeper glacial deposits provided the parent materials for soils that formed at the surface during Pleistocene times. Those soils were the initial tools that researchers used more than a hundred years ago to identify different till layers in the thick package of glacial deposits across the state, which in portions of southwest Iowa are at least 350 feet thick. A zone with soil properties such as a distinctive color, soil structure, subtle horizons, a spike in clay or organic carbon, and even wood fragments or charcoal is a kind of geologic marker—a layer that can be traced over horizontal distances and that signifies a time of relative landscape stability when deposition and erosion were minor. It also implies a major difference in the origin and age of the geologic materials above and below it and thus in the ice ages that those layers represent.

Once the glacial till layers had been broadly divided, researchers further distinguished them on the basis of texture and the mineral composition of their pebbles, sand, silt, and clay. Initially, geologists separated the thick deposits left by glaciers more than half a million years ago into two layers, representing two ice ages—previously called the Nebraskan and Kansan glaciations. Then in the late 1970s, they uncovered evidence for more than two till layers in southern Iowa. Several test holes, including one drilled to a depth of 190 feet in Union County south of the town of Afton, showed that the thicker and older "Nebraskan" till actually consisted of three different till layers separated by paleosols. In addition, researchers found even more till layers deeper in the cores (Boellstorff 1978). In the early 2000s, cores taken near the same location confirmed the presence of the multiple tills and the paleosols that separated them and afforded geologists an opportunity for detailed study and preliminary dating of the layers.

Determining the ages of these very old tills and their corresponding glacial advances into the midcontinent has been a challenge, but the

Afton cores finally made progress possible. An exciting early discovery in some of the cores' interglacial sediments was the presence of datable volcanic ash from three ancient Yellowstone eruptions previously shown to have occurred 630,000, 1.3 million, and 2.1 million years ago. Another fascinating and very fortuitous property of these sediments was their magnetic polarity. Earth's north and south magnetic poles slowly move over geologic time—every several hundred thousand years, they switch places (so beware, geocachers). Sediments carry a paleomagnetic signature indicating whether they were laid down during a period of normal polarity—when the magnetic north pole is in the north, like the present time—or during a period of reversed polarity. Using this information, geologists were able to place the southern Iowa tills into rough age brackets by comparing their paleomagnetic signatures to those of marine sediments, which had been dated previously by other means (Roy et al. 2004).

Coming up with more precise ages has been difficult. Even if charcoal is present, a sample older than about 50,000 years is too old for radiocarbon dating because it has lost nearly all its carbon-14, an isotope with a relatively short half-life. A couple of other dating techniques are good only to about 200,000 years ago. More recently, however, geologists have come up with an intriguing dating method involving paleosols and the ratios of two radioactive isotopes they contain: beryllium-10 and aluminum-26, which are produced by cosmic rays hitting an exposed soil surface. Based on the decay half-lives of the isotopes, researchers are able to determine when a particular paleosol was buried by deposits thick enough to block cosmic rays from reaching it. In this way, the burial date gives a fairly precise age for deposition of the glacial till now overlying the paleosol. This cosmogenic dating of paleosols has made it possible to begin to establish the timing of major advances of the Laurentide ice sheet over the past 2.7 million years (Balco and Rovey 2010; Rovey and McLouth 2015).

There is now general consensus that our little part of the continent experienced at least seven major periods of Pleistocene glaciation prior to 500,000 years ago. In Iowa, we currently label them simply as Pre-Illinoian in age, pending further research and dating. Figure 34 shows

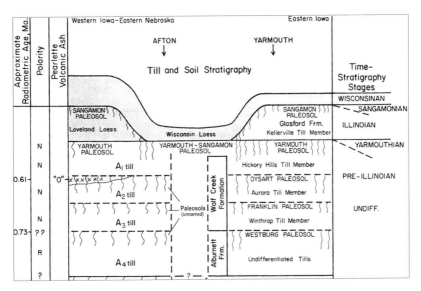

34. Upper portion of the Pleistocene stratigraphy across southern Iowa. Adapted from Timothy J. Kemmis, E. Arthur Bettis III, and George R. Hallberg, "Quaternary Geology of Conklin Quarry."

only the upper four of the seven Pre-Illinoian tills identified in southwest Iowa and the paleosols formed in them. It also illustrates how the units probably correlate with the four tills and paleosols found in southeast Iowa.

Although not as well studied and not as complete a record, Pre-Illinoian glacial deposits exist in most of the rest of the state as well. Iowa also has an area of younger Illinoian deposits in seven southeast counties bordering the Mississippi River, and geologists with the Iowa Geological Survey are exploring and mapping a third Wisconsinan till in north-central Iowa (Kerr et al. 2019). At latest count, our state has glacial deposits from at least eleven ice ages, which means that Iowa possibly has the most complete land record of Pleistocene glaciation anywhere on Earth. That is certainly true in North America, where some of the ice sheets never reached farther south than Iowa, and subsequent ice sheets erased deposits from some or most of the older glacial advances to our north. Like the tale of the three bears, Iowa's geographic position appears to have been just right.

From studies of cores, it is clear that there were active landscape processes happening between the ice ages. During interglacial periods when the surface was free of ice, life once again flourished, and soil formation began anew. Although glacial till and meltwater deposits make up the bulk of the Pleistocene sediment package, intervening layers of windblown loess, lake silts and clays, and stream deposits are commonly seen in quarry exposures and drill cores as well. Recognition of loess deposits between till layers was slow in coming, but loess is now a preferred explanation for the thick layers of fine-grained homogeneous materials—containing no pebbles and very little sand—often found between till layers. In some of the Pleistocene paleosols found in southern Iowa, the upper horizons formed in these loess parent materials, while the lower horizons formed in the underlying glacial till.

The thickest of the Pleistocene paleosols is the Yarmouth-Sangamon paleosol, which developed over some 200,000 years. It lay at the land surface through two interglacial stages—the Yarmouth and the Sangamon—and two glacial stages—the Illinoian and most of the Wisconsinan—before Late Wisconsinan loess buried it and halted soil-forming processes. The paleosol's parent materials included the youngest Pre-Illinoian till, overlain by at least two loess deposits of Illinoian age or older (Woida and Thompson 1993). But unlike the loess of the Peoria Formation, these loess deposits fell slowly enough that soil formation was able to keep pace by incorporating the new material into the soil profile as it built upward, eventually becoming more than 20 feet thick. (In eastern and western Iowa, the Yarmouth and Sangamon paleosols are separated by Illinoian till or Loveland loess.) Some of the next-generation soils described previously, such as the Clarinda and Bucknell soils, are rooted in the Yarmouth-Sangamon paleosol. Other Clarinda and Bucknell soils found at lower elevations formed in the older, thinner, unnamed Pre-Illinoian paleosols that erosion had exposed on hillslopes (Woida and Lensch 2015).

There is one thing we can be certain of: after the deposition of hundreds of feet of glacial sediments during Pre-Illinoian times, there was a lot of erosion happening on the landscape once the final glacier retreated. In the mid-twentieth century, the savvy and persistent Robert Ruhe and a colleague studied the many paleosols exposed in fresh roadcuts and railroad cuts along a 60-mile transect in Adair, Cass, and Pottawattamie Counties. From their detailed observations, they concluded that southern Iowa has a stepped landscape formed through a sequence of erosional episodes alternating with periods of stability during the last few hundred thousand years (Ruhe and Cady 1967). This was a pivotal contribution to geologists' understanding of landscape development. Ruhe's stepped landscape model and his work in other parts of the U.S. led to an important new international field of study called soil geomorphology (from the Greek *geo*, referring to earth, and *morph*, meaning "shape").

In general, the Pleistocene record is much less complete in other parts of the state than in southern Iowa, probably the result of more extensive erosion that must have occurred both during and after Pre-Illinoian times. It is difficult, however, to identify negative evidence—in other words, evidence of a missing layer removed by erosion. Where present, though, paleosols can be very helpful because a buried B horizon with no sign of an A horizon implies erosion. Remember, soils always form from the top down, beginning with the A horizon.

We do have plenty of evidence for erosion during the Wisconsinan glaciation in parts of Iowa not covered by ice at the time, particularly parts of the Iowan Surface in the northeast quarter of the state—see figure 25 in chapter 3. Drill cores obtained there from paha looming as much as 100 feet above the surrounding landscape often exhibit the same geologic layers and paleosols that we see in southern Iowa. The paha are elongated loess-capped hills unique to northeast Iowa that are believed to be remnants of a former and higher landscape. If that's correct, it would imply that strong winds and meltwater generated by the nearby Des Moines Lobe ice removed as much as 100 feet of Pre-Illinoian till and the paleosols along with it. Now buried by younger sediments, this erosional surface as seen in exposures is often marked by a very thin horizontal stone line or stone zone. In other places, large boulders known as glacial erratics lay strewn across the landscape.

In both cases, the stones are the heavy stuff, the lag left behind when wind or water winnowed out and carried off the lighter particles. A similar process probably occurred west of the Des Moines Lobe, although not as extreme. Jean Prior's landforms book gives a thorough description of the visible and near-surface landscape features across Iowa that resulted from the complex progression of deposition, erosion, and landscape stability since Pre-Illinoian times.

Buried paleosols are the best evidence we have of the landscape processes and paleoenvironments that existed in Iowa when it was ice-free during the last couple of million years. Soil formation ceased when these soils were later buried, leaving profiles with clues to ancient environments in the form of soil properties. Many of them carry implications similar to those same properties observed in surface soils in Iowa or elsewhere (see chapters 1 and 3).

For example, granular soil structure in a paleosol indicates a former A horizon even if most or all of the organic carbon is gone. If it is several inches thick, it likely formed under grassland. Abundant clay coatings on ped surfaces suggest long-term landscape stability and soil formation under subhumid conditions, which promoted leaching of calcium carbonate and downward translocation of clay. A siltier zone with subtle platy structure found directly above a Bt horizon may be an E horizon that formed under woodland vegetation.

The depth to pedogenic carbonates in a paleosol can suggest mean annual precipitation rates in the past, and the presence of other minerals such as barite and gibbsite provides clues to climate or topographic position. Some clay minerals point to certain parent materials, while other clays suggest length of time and weathering at the surface or position on the landscape. Slickensides indicate pronounced shrinking and swelling, likely caused by seasonal moisture deficits a few months long—an environment like that in eastern Australia or parts of the eastern half of Texas, more suited to grasses than trees. As already explained, stone lines and missing A horizons indicate significant erosion.

Many of these features are meaningful clues only when they occur together with other properties. Some, such as orange or black redox

35. Left: Armstrong soil profile, Marshall County, photograph by Gerald Miller. Right: Microscopic view of the reddish B horizon of a Sangamon paleosol, photograph by the author.

features and gray depletion features, are not usually a reliable clue to the past at all, because iron and manganese oxides usually continue to move through the paleosol profile with groundwater long after burial. However, certain paleosols found across much of the state often have a characteristic reddish color throughout the matrix. Referred to as Sangamon paleosols, they formed over a period of some 70,000 to 90,000 years during the interglacial period between the Illinoian and Wisconsinan ice ages in the many different parent materials at the surface (see fig. 34). Where these paleosols are red, their color, mineralogy, and chemistry suggest that they formed when the climate was warm to hot, with long periods of drought (Markewich et al. 2011). Late Wisconsinan loess eventually buried the Sangamon paleosols. But after erosion had removed much of the loess, soil formation began anew and resulted in modern soils like those of the Armstrong series and the Adair series, which have very complex histories.

For example, the profile in figure 35 of an Armstrong soil from Marshall County has a dark A horizon developed in loess. The loess overlies a reddish Sangamon paleosol, which formed much earlier in loamy erosional sediments that now comprise the 2Bt1 horizon and into the B horizon of an even older Pre-Illinoian paleosol—now the 3Bt2 horizon—that had been previously exposed on one of Ruhe's erosion steps. Armstrong soils are Alfisols, so an E or a BE horizon is sometimes present. The accompanying photomicrograph shows the Sangamon paleosol's pronounced blocky structure, clay coatings that developed under a subhumid climate, and extensive iron staining from a subsequent long and hot dry period unlike anything experienced in Iowa during historic times.

There are many, many more examples. I've barely scratched the surface of a complex subject, but I think you can see that deciphering paleoenvironmental conditions from these old paleosols is a challenging business. A soil that lay at the surface for such a long period of time is truly a palimpsest: an intricate, layered record of changing conditions of climate, vegetation, and landscape position, all of which left their imprint on the soil's physical and chemical properties.

The Archaeologist's Tool

Compared to Pleistocene paleosols, most buried Holocene paleosols in Iowa lay at the surface for only a few hundred years or less. Nevertheless, they too contain clues to what was happening on the ground at the time of their formation. Many of them consist solely of a dark A horizon, but its degree of soil development—for instance, the strength of its soil structure or the abundance of its coatings—can suggest its length of time at the surface. Seeds, pollen, and other plant parts as well as insects, snails, and the remains of small rodents, if found, are excellent indicators of the climate and flora and fauna of the Holocene ecosystem. Best of all, if a Holocene paleosol contains charcoal or other organic fragments, it is young enough to be radiocarbon-dated!

For archaeologists, the very presence of a buried soil in a column of sediment layers is extremely useful because it indicates where a ground surface remained unburied long enough for a soil to form. Any such soil could potentially contain evidence of human habitation such as charcoal, flint, or other artifacts, so identifying a paleosol is a crucial first step toward discovering evidence of human occupation (fig. 36).

Archaeological sites in Iowa are concentrated in valleys, where soils form on floodplains, terraces, and alluvial fans (Alex 2000). Early humans typically settled in valleys where food and water and

36. Paleosol with white flint chips, buried by postsettlement alluvium, in an Allamakee County valley. Photograph by Richard Rogers, USDA–Natural Resources Conservation Service.

a transportation corridor were readily available. They were especially drawn to alluvial fans, which are roughly triangular landforms that develop where tributaries enter into a larger valley and drop their sediments. Both fans and terraces sat above the floodplains and so provided protection from seasonally high water levels in the rivers. Fringe areas around wetlands were popular real estate because they were close to an abundance of edible plants (Holliday 1992). Floodplains and the wetlands themselves, however, were seldom occupied due to damp ground. As one archaeologist put it to me, the goal was to have a wet bottom topographically speaking and a dry bottom anatomically speaking.

Any given valley may contain more than one Holocene paleosol, each buried by a deposit of alluvium—or it may have no paleosol at all. Where potential buried soils are hidden from view in thick deposits of alluvium, archaeologists must work closely with geologists or soil scientists. Knowledge of how alluvial deposits are distributed in a valley allows archaeologists to concentrate their arduous sampling efforts where they will be most effective.

Repeated episodes of downcutting during the Holocene Epoch deepened many Iowa valleys, and subsequent deposition partially filled them in again. In the 1980s, Art Bettis—an archaeologist, soil scientist, *and* geologist—and Dean Thompson established that three major episodes of Holocene alluvial deposition and downcutting took place in Iowa and parts of some adjacent states. Based on numerous radiocarbon dates, the first and thickest alluvium was deposited, then downcut by stream erosion, prior to 4,000 years ago. The second episode of deposition and stream incision occurred between about 4,000 and 200 years ago. The last major cycle of deposition began with the Euro-American settlement that brought the rapid spread of agriculture to what is now Iowa. The top geologic layer in nearly all Iowa valleys consists of a few to several feet of postsettlement alluvium: the deposits that resulted from the massive erosion of loess soils disturbed by intensive cultivation after settlement. Soils that formed in this geologically young alluvial layer found on floodplains and terraces are typically Entisols.

A Lesson for Today

One of the most remarkable things I discovered from studying Iowa's ancient paleosols is the amazing resilience and persistence of certain soil features, despite their being overridden and consolidated by one or more glaciers with ice thousands of feet thick. As long as an overlying deposit such as loess or lake sediments protected the upper horizons from erosion, the original granular peds often survived in these ancient soils even though the soil pores understandably did not (fig. 37). Surprisingly, thin sections from my studies of buried soils that formed during the last three Pleistocene interglacial periods revealed granular structure in the upper several inches in eight of fourteen paleosol cores from Madison, Adams, and Clarke Counties. Also preserved were earthworm casts, worm tubes, and burrows called krotovina—

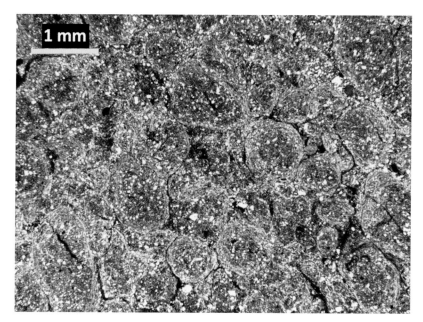

37. The A horizon of a 600,000-year-old paleosol between the A1 and A2 tills in Adams County. Circularly polarized light reveals orange circles where clay was smeared onto granular peds in the digestive tracts of earthworms. Photograph by the author.

yet another vestige of Russian terminology—from soil fauna such as crayfish or small mammals.

In contrast, modern topsoils subjected to tillage year after year typically display no soil structure at all. As a result, despite existing at the very surface, they have very little of the porosity needed for rainwater to infiltrate into the ground (see fig. 45, chap. 6). It is largely because of this degradation of soil structure that runoff of rainwater carrying sediment and pollutants to our streams and lakes is so rampant in Iowa.

To begin to understand how we got from the strong, persistent morphologies preserved in the soils of the past to the impoverished and compacted topsoils of today, we return now to a discussion of modern soils on our present landscapes. Formation of our surface soils continues today and will continue into the future until such time as they are eroded off or buried. So the story of these soils is still unfolding, and for better or worse, we humans will be the critical factor in how that story unfolds—the subject matter for the second part of this book. It begins with the early days in Iowa's history as a state, a time when settlers from eastern and southeastern states arrived and began to transform Iowa's natural landscapes and pristine soils at a pace the land had never before experienced.

PART TWO

The Sixth Factor

People, Agriculture, and Soils

CHAPTER FIVE

Reaping the Bounty

So completely has the whole State passed beneath the plow,
so quickly assumed the appearance of one vast farm, that one
who studies the Iowa of today realizes with difficulty the
strange picturesque wildness of fifty or sixty years ago.
—THOMAS MACBRIDE, 1895

IN THE GEOLOGIC PAST, it typically took Iowa's soils centuries to respond to fluctuations in climate and the shifts in flora and fauna that followed those changes. In comparison, soils change very rapidly in response to human activities. As will become clear in chapter 6, soil formation under human occupation is often a case of negative formation, with the A horizon being especially damaged by erosion and degradation (Sandor, Burras, and Thompson 2005). Some human-induced changes—fertilization, tillage, and so on—are intentional but others, such as erosion, are unintended. Regardless, the scale of anthropogenic changes can be huge, and they have increased exponentially since the nineteenth century. Even though we humans technically are one element of the soil-forming factor called organisms, many researchers consider human impacts on the soil to be a sixth factor because of the extraordinary rate at which they transform soil properties. Unlike the other five factors, which act over centuries or millennia, this sixth factor operates on a timescale of decades (Richter 2007).

The online World Reference Base for Soil Resources includes as one of its thirty-two major soil groups the Anthrosols—soils formed or profoundly modified by human activities, including burials, partial removal, cutting and filling, waste disposal, the application of manure, cultivation, and irrigation (Krasilnikov and Arnold 2009b). Anthrosols mainly are mapped in northwest Europe, where the addition of manure and animal bedding since medieval times has thickened soil profiles by as much as 2 to 3 feet; in Southeast and East Asia, from intentional puddling of rice fields; and in the Middle East, from centuries of irrigation with sediment-laden water. There are small areas in the United States that fit the bill, including river terraces in Maryland, where deep A horizons grew upward in layers of food refuse from early Native American habitation, mainly oyster shells and fishbones.

Although intensive agriculture hasn't been around long enough in Iowa for any of our soils to qualify as Anthrosols, it has wrought changes on a comparable scale in much less time (Veenstra 2010). Under intense management, soil changes can occur at depths up to 5 feet (Burras 2016). Just recall the soil order known as Entisols from chapter 2—soils consisting simply of a thin A horizon over the C horizon. Most of the Entisols on the steep loess slopes of western Iowa now exist where there were once Mollisols with thick rich A horizons and mature B horizons, both since lost to erosion. Likewise, many urban areas have Entisols developed in fill dirt that buried or replaced the original soils or in the C horizon of soils stripped of the horizons above it.

These changes have been enormous, and they have advanced at an unparalleled rate. A landscape that before 1850 flaunted more than 23 million acres of prairie—with its incomparable soils—had lost most of it by 1930. In the words of Larry Stone, "this astounding transformation from a natural landscape of wild places . . . to a checkerboard of manicured crop fields, cities and roads took place in barely 60–70 years, less than a lifetime" (2000). In 2015, the Iowa Department of Natural Resources estimated that Iowa had only about 30,000 acres of native prairie left, roughly one-tenth of a percent of the original prairie ground.

Over a span of about 200 years, our soils changed radically as agriculture evolved from a sparse patchwork of small farms cultivated by animal and human labor to a landscape of fewer and larger farms sustained by high-tech equipment and billions of pounds of agrochemicals

(Mutel 2008). The human story of how Iowa was transformed from the land of lush prairie to the land of tall corn is a tale of hope, pioneer fortitude, spectacular success, and record-breaking production but also at times one of heartbreaking disappointment and bankruptcy. Of course, ultimately, the most lasting legacy was the destruction of many of Iowa's natural ecosystems.

For the Native Americans who populated the land that came to be called Iowa, the story is one of loss and change almost beyond imagination. Agriculture was a critical part of life for Iowa's Indian tribes—the Ioway, Sauk, Meskwaki, Ho-Chunk, Potawatomi, and others. The soil was not only the source of much of their food, it was the material from which they shaped their household vessels and which served as the mortar in their sod homes. The most important Native American crops were corn, beans, and squash, usually interplanted in what has come to be called a three sisters arrangement—an ingenious system that provided a trellis for the beans, nitrogen fixation for the corn, and ground cover to stifle weeds and retain moisture. Other crops included pumpkins, sunflowers, melons, and tobacco. The planting sticks and shoulder blade bones of bison that Native Americans used to shape small mounds for the large seeds were ineffective tools in the tough prairie sod, so they grew their crops on floodplains and low stream terraces and in woodland clearings. Some tribes hunted local bison and other game, while others followed bison herds throughout the year. It was a system of agriculture adapted to the limits of the resources available (Lengnick 2015). Although archaeologists have documented some soil erosion, it was much less than what occurred in Asia, Europe, and northern Africa, where small-grain crops such as wheat, rye, and barley usually involved field-scale tillage to create a fine seedbed (Magdoff and Van Es 2009; McNeill and Winiwarter 2004).

Euro-American Settlement: The 1800s

Between 1824 and 1853, the U.S. government's Bureau of Indian Affairs negotiated nine treaties with Iowa's tribes. Over that time, the government paid a total of $2.9 million for the entirety of the land of Iowa, which averaged out to 8 cents per acre. (In 1856, the Meskwaki were

allowed to buy back 80 acres of land in Tama County at the enormously
inflated price of $12.50 per acre.) Before Iowa became a state, Congress
set up land offices to parcel out acreage to Euro-American settlers,
with the first two offices being established in Dubuque and Burlington
in 1838. In the preceding five years, 23,000 squatters had already staked
out claims; they then had to compete to legally purchase the land they
had claimed at a minimum price of $2.50 per acre (Wall 1978).

The settlers' first choice of land was in wooded areas. Most came from
Ohio, Indiana, Pennsylvania, and New York—with a smaller number
from Kentucky, Missouri, Virginia, North Carolina, and Tennessee—
where trees provided plentiful wood for building, heating, and cook-
ing. Others came directly from European countries such as Sweden
and Holland. Seeing the great green sea of treeless prairie, they assumed that if it was too poor to grow trees, it was too poor to grow
crops. As late as 1847, prairie ground was selling for $3 to $10 per acre,
while timberland cost $30 to $50 per acre (Thompson 2014a). But those
who had stopped over for a year or two at places along the way, such
as Illinois, had some experience coping with the prairie, and others
who bought prairie eventually found that the land offered deep fertile
soil once they broke the sod (Schwieder 1996; Schwieder, Morain, and
Nielsen 2002).

To accomplish that, many pioneers turned to professional prairie
breakers—sodbusters—who lived in covered wagons and arrived with
three to eight yoke of oxen and a breaking plow, a gigantic wooden
moldboard plow covered by sheet iron that could plow 2 acres in a
day. However, the service often cost more per acre than the land itself.
In 1837, John Deere invented his steel prairie plow, which was drawn
through the sod by a team of three horses and steered by the farmer
on foot (fig. 38). It consisted of a blade—called the plowshare—which
cut the soil into slices, and the moldboard, which inverted the furrow
slice to bury the sod. It was a revolutionary invention, for a farmer
could turn 80 acres of prairie and have it planted in just two months
(Wall 1978). By then, traditional hill crops had long disappeared, replaced by row crops more amenable to tillage. Regardless of which
plow was used, it worked by inverting the sod, over time killing the

38. Breaking prairie in Sioux County with moldboard plows, circa 1890.
Courtesy of the Ireton Area Historical Society.

prairie vegetation and leaving the soil uncovered for much of the year.
It wouldn't be until the 1950s that farmers had another choice in tillage tools.

Initially, corn was the pioneers' grain of choice for human and livestock consumption, and potatoes were a staple at most farm family
meals. Wheat, beef, and pork were the chief commercial products; during the Civil War, Iowa ranked second nationally in wheat production.
But inverting the prairie sod had forever disrupted the ecological balance of nature acquired over several millennia. While the native prairie was largely resistant to disease, the imported cultivated plants of
wheat, oats, and flax were vulnerable to rust fungi, and orchard trees
were vulnerable to scab and blight. Only Indian corn seemed to be
resistant. With the ecological balance of the insect community also
radically transformed, grasshoppers ravaged grain crops three out of
four years in the mid-1870s (Wall 1978). These were hard times on the
frontier, when settlers faced food and feed shortages, terrifying grass
fires, bitterly cold winters with blinding snowstorms, the tragic loss

of young children to illness, and long-term or permanent separation from loved ones back home.

After the Civil War, farmers were urged to diversify, and following the locust scourge many began to grow more corn and less grain. By 1880, most families had turned from subsistence farming to commercial farming, aided by the laying of five major railroads with connections to eastern markets. Corn soon became the state's major commodity, with Iowa bypassing Illinois to become first in the nation in corn production in the 1880s. At first, farmers used the corn mainly to fatten hogs, which eventually gave birth to the Iowa swine industry. Grazing of beef cattle expanded considerably, as the invention of refrigerated railcars promoted the spread of cattle-feeding facilities, slaughterhouses, and packing and distribution plants (Lengnick 2015). In these days before the tractor, farmers also grew large quantities of oats to feed their horses, of which Iowa had many, second in number only to Texas. Regional agricultural specialties of the day included dairy in northeast Iowa, grass-seed production in southwest Iowa, flax in north-central Iowa, and popcorn in Sac and Ida Counties (Schwieder 1996).

The state had been settled in a series of frontier zones, beginning in the southeast and culminating in treeless northwest Iowa in the 1860s and 1870s, where settlers constructed and lived for many years in sod houses built from the plow's furrow slices. By the 1880s, the Iowa frontier was a thing of the past, and Iowa had undergone a phenomenal change. Between the statehood year of 1846 and 1850, only five years later, the state's population had doubled. By 1900, it had grown to 2.2 million—an elevenfold increase—and 95 percent of the prairie had been converted to agriculture. Four million acres of woodland were also lost to either agriculture or timber cut for buildings, railroad ties, and mine shafts (Wall 1978). Steam shovels and the manufacture of drainage tile made it possible to begin draining and converting wetlands to agriculture (Doak 2015). The channelization—the straightening and deepening of rivers and drainage ditches—that had begun in the late 1800s peaked in the early 1900s with heavy equipment and eventually initiated cycles of erosion that deepened and widened many streams (Stone 2000). More than a hundred years later, historic channelization continues to negatively affect many Iowa watersheds.

39. Installing drainage tile in Kossuth County, circa 1912. From the William Shirley Collection, State Historical Society of Iowa, Des Moines.

Agricultural Expansion: 1900 to the 1950s

The period between 1909 and 1914 has been called the golden age of American agriculture, and it was indeed a boon time for Iowa. Nine percent of the nation's gross farm income came from Iowa, and more than 50 percent of livestock receipts at the Chicago stockyards originated in Iowa (Schwieder 1996). By 1910, land values in the state had risen to more than $80 per acre. More land became available for cultivation as a result of a 1908 change to the state constitution, which authorized the formation of drainage districts and the use of eminent domain to dredge wetlands and construct drainage ditches and levees (fig. 39). By 1930, 95 percent of the wetlands in north-central Iowa had been converted to farmland.

Corn and hogs dominated farming in the early 1900s. A traveling exhibit called the Seed Corn Gospel Train visited ninety-seven Iowa counties, training farmers on the best ways to select seed corn, rotate crops, handle manure, and raise hogs. The Iowa General Assembly passed the Agricultural Extension Act in 1906, which preceded the national cooperative extension program by eight years. It made state funds available for farming demonstration projects and eventually led to the creation of county extension offices. A few years later, the Iowa Agricultural Experiment Station, a research program established in 1888, began work on soybeans by testing 3,000 strains from China. Soon soybean crops took root, initially as forage crops for livestock (Wall 1978).

During World War I, Iowa farmers supported the army and the Allies overseas not only through food production but through food conservation, as county extension agents held demonstrations for farm women on gardening, canning, and raising poultry. As a result of the war effort, farm production records were set and annual sales exceeded prewar production by 26 percent. However, the government ended price guarantees for crops and livestock in 1920, allowing supply and demand to once again dictate prices. By 1921, with export markets dwindling, large surpluses caused farm prices to fall by two-thirds. Times became lean once again for Iowa's farmers. Many had invested in more farmland, but with the collapse of prices the land bubble burst and the market value of land fell below what was owed on it. For Iowa's farmers and the many small towns that depended on agriculture, economic depression set in several years before the stock market crash of 1929 that led to the Great Depression.

The Depression of the 1930s affected all Americans, but farmers on the Great Plains suffered a double disaster as a series of historic droughts ruined crops and grazing lands. The decade started with dry years in 1930 and 1931, and extreme drought followed in 1934, 1936, 1939, and 1940. Paradoxically, Iowa farmers harvested their largest corn crop ever in 1932, but low prices—less than half the average prices of the 1920s—nevertheless caused economic ruin for some. The 1936 drought, which followed an exceptionally long and cold winter, was especially devastating (Swaim 1986). Although Iowa was not as hard hit by the Dust Bowl as Kansas and Oklahoma, winds in the mid-thirties darkened skies, piled dust 2 to 3 feet high around fences and buildings, and settled so thickly on pastures that cattle would not eat and "wandered about bawling their hunger," wrote Iowa poet James Hearst (2001). Grasshoppers thrived in the hot dry weather and destroyed any surviving crops. The harsh conditions forced some farmers to give up and join the exodus to the West.

Soil erosion on fields stripped of vegetation during these droughts was astounding. Based on a nationwide reconnaissance erosion survey of every county in the nation conducted in 1934 (McCormack and Paschall 1982; Soil Conservation Service 1935), the Natural Resources Conservation Service later estimated that land in row crops and small grains in six sample Iowa counties in 1930—not even the worst of the drought years—was losing soil to erosion at an average rate of 15 tons from every acre (Argabright et al. 1995). For comparison, the rate in 2017 for Iowa as a whole was nearly 6 tons per acre, still the highest in the country, as you will see in chapter 6.

Ironically, the 1930s also brought progress: tractors with rubber wheels, combines to replace threshing machines, electricity, and indoor plumbing. In 1939, Iowa had more tractors than any other state, and half the corn crop was picked mechanically. Besides speeding up production, the proliferation of tractors in the twenties and thirties increased production by freeing up considerable acreage previously used to grow feed for horses.

Henry A. Wallace, born in Adair County, had begun experimenting in the late 1920s with corn hybrids to produce plants with longer ears, larger kernels, and greater resistance to disease. The first hybrids from Iowa State College (renamed Iowa State University in 1959) went on the market in 1932, and by 1939 three-fourths of Iowa's corn acreage was planted in hybrids. By 1944 it was all in hybrids, compared to less than 60 percent nationwide (Wall 1978). With companies like Wallace's Pioneer Hi-Bred Corn Company and DeKalb AgResearch always developing and selling new hybrids, production across the entire Corn Belt increased as yields jumped from 22 bushels per acre in the early thirties to 32 bushels per acre in the early forties. (By that time, Wallace had already served as U.S. secretary of agriculture and had become vice president under Franklin Roosevelt.)

The 1940s saw the end of the horse-drawn era and greater demand for farm products, largely driven by the economic stimulus of World War II. Spurred by the U.S. Department of Agriculture's motto during the war—"Food will win the war and write the peace"—Iowa's farmers set production records every year from 1941 to 1945. Modern pesticides had greatly reduced crop loss, and hybrid corn with its stronger stalks had made the use of combines more feasible. Given the increased international demand for agricultural products, there was a push to produce more with less labor. So farmers shifted from human and animal labor to machines whenever possible, typically sharing harvesters and other machinery with neighboring farmers.

Agricultural Transformation: 1950s to the Present

By the middle of the twentieth century, Iowa agriculture had experienced a massive transformation. Animal power had given way to fossil fuels, pastured livestock to an increase in animal confinement operations, and a regional market to an international market. Intensive cultivation of hybrid corn had created more need for pesticides and drained the soil of nitrogen, which together with topsoil loss from erosion increased the need for chemical fertilizers. Inorganic nitrogen fertilizer became readily available in the United States in the 1950s, when pipelines began to carry liquid ammonia north to the Corn Belt from natural gas plants in Texas, Louisiana, and Oklahoma.

With the further mechanization of farming along with frequent droughts in the 1950s, the farming population began to decline, epitomized by the closing of hundreds of one-room schoolhouses (Jeff Bremer, personal communication). Up to four consecutive years of drought in some parts of the state kept production down despite the increasing application of nitrogen to fields. Experiencing high rates of wind erosion during drought years, some farmers began using the chisel plow instead of the moldboard plow. The chisel plow, which left more plant residues on the soil surface, had been invented by an Oklahoma farmer during the Dust Bowl years in a desperate attempt to save his soil from the wind.

Most Iowa farmers recovered in the sixties and began moving toward greater specialization and a stronger commercial orientation.

By 1970, the state's corn production exceeded a billion bushels, more than double what it had been forty years earlier. At the same time, the 1950s and 1960s saw an enormous increase in global agricultural production known as the Green Revolution, when wheat and rice yields in countries like India and Pakistan improved dramatically. A major figure in the movement was Norman Borlaug, who was born and raised in Cresco, Iowa. He eventually won the Nobel Peace Prize for his role in genetically engineering high-yielding and disease-resistant cereal grain varieties and for promoting the mechanization of cultivation and the use of chemical fertilizers and irrigation in South Asia and elsewhere. It was a truly revolutionary initiative to address the global food supply in the face of exponential population growth, and many have credited it with saving as many as a billion lives. Others have criticized it for bringing large-scale monoculture to lands, soils, and cultures incapable of supporting it, so that off-farm inputs are intensive and reap big profits for U.S. agrochemical corporations.

Back at home in Iowa, crop prices in the early 1970s were high, the surpluses were gone in the wake of tremendous exports, and the USDA urged farmers to grow more. The prosperity of the seventies, including the bonanza years of 1973 and 1974, led many farmers to buy more land. Banks had encouraged them to borrow more money to invest in Iowa's black gold and to buy more and bigger machinery to farm it. As a result, the size of farms increased and the number of farms decreased. With fewer families farming the land, rural population numbers continued the decline that had begun in the 1950s. Iowa's tapestry of abundant small-scale diversified family farms became one of fewer small farms and more large-scale specialized corporate farms and feedlots. In hindsight, from today's more informed perspective, these changes reinforced a trend that has not only transformed rural Iowa but has caused major damage to our soil and water.

Beginning in the late 1970s, when most of the country was economically stable, a crippling farm crisis rapidly accelerated the decline in the number of family farms. As farm debt soared due to unchecked land and equipment purchases, record production combined with high interest rates and oil prices led to a sudden collapse of land and commodity prices for U.S. farmers. In 1981, interest rates on farm loans peaked at 21.5 percent, and total interest payments alone exceeded the year's net income for some farmers. Midwest droughts in 1983 and 1988 further damaged farm income at the worst possible time. At the epicenter was Iowa, where a record number of farmers, unable to meet their financial obligations, lost their farms. Protest rallies in support of neighboring families frequently accompanied farm foreclosures. The frequent farm auctions were heartbreaking affairs for those who had lost their farms, often as a result of factors beyond their control.

Anxiety, depression, and the suicide rate among farmers soared. The Farm Crisis was the worst time Iowa farmers had experienced since the Depression, and its consequences extended far beyond the farm (see the 2013 Iowa PBS documentary *The Farm Crisis*). Many of the small-town banks that had lent farmers money were forced to file for bankruptcy, and many small-town merchants closed their doors for lack of business. More than 140,000 people moved off Iowa farms during the 1980s. As rural counties and small towns lost population, their schools were forced to consolidate into larger districts, a trend that has continued into the twenty-first century. Between 1950 and 1992, the number of farms in Iowa decreased by half, from about 200,000 to fewer than 100,000, and the average farm size nearly doubled, from 170 to 325 acres (Schwieder, Morain, and Nielsen 2002). By 2017, there were only 86,100 farms in the state.

Throughout this dynamic, sometimes turbulent history of Iowa agriculture and despite short-lived ups and downs, the overall trend has been one of steadily increasing production. As an example, the corn harvest rose from 2.3 billion bushels in 2007 to 2.6 billion bushels in 2017, a 13 percent increase in just ten years. During that same period, soybean production went from 431 million bushels to 553 million bushels, a 28 percent increase. However, fewer family farms benefited from the bounty because the number of small farms had continued to fall, while farms larger than a thousand acres had increased by 13 percent—to 8,417 farms—in that ten-year period.

In many ways, today's farming bears little resemblance to that of sixty or seventy years ago. Planting of genetically modified seed—

80 percent of it controlled by Monsanto (now Bayer)—is done with enclosed-cab diesel tractors equipped with precision GPS and capable of covering many acres in an hour. Crops can be harvested with sixteen-row combines and loaded onto grain carts that lumber across the soil surface with axle loads as great as 20 or 30 tons—two to three times what is allowed on public roads, with the exception of farm vehicles. Ubiquitous fertilization with inorganic nitrogen has decreased the need for proper crop rotation, so that most farmers now use, at best, a simple three-year "rotation" of corn-corn-beans. Pesticides leave fields weed-free, neat, and clean, but the soils ailing. However, there is some good news. Although chisel plows, heavy discs, and even a few moldboard plows are still commonly used to break up the compacted soils, more and more farmers have been practicing various forms of conservation tillage since the 1980s (see chapter 8).

From this look back over nearly two centuries of Iowa history, we can see how settlers of European origin eventually transformed the land between the Missouri and Mississippi Rivers into one of the premier agricultural states in the country. Agricultural production has contributed immeasurable wealth to Iowa's economy and the well-being of its citizens. At the same time, many now realize that this abundance has come with a price—in the form of factory farms, polluted water and air, the decline of family farms and small towns, food that is less healthful, and chronic illnesses in rural areas. Often overlooked, however, is the toll it has exacted on our soils, which of course are the ultimate source of this bounty. Throughout our history, they have been the substrate, the food supply, and the water-delivery system for our crops, our woodlands, our gardens and lawns, and our livestock. But human activities—the sixth factor of soil formation—have radically changed this life-giving source. Drained, pulverized by tillage, compacted by heavy machinery, permeated with synthetic chemicals, left bare and exposed to the elements for seven months out of the year, overgrazed, stripped, or buried by concrete or pavement—all these have left an enormous mark on the physical, chemical, and biological quality and health of our soils. In addition to the steadily decreasing natural fertility and declining tilth of the soils themselves, erosion of this precious resource is contributing a shocking amount of nitrogen, phosphorus, and sediment to our rivers and lakes and to our downstream neighbors.

Fortunately, many Iowans are beginning to recognize the limits of the land. Because its soils were so rich and plentiful and its resources seemed inexhaustible, most people didn't recognize those thresholds for much of Iowa's history. Compared to the forest soils of the eastern United States, which degraded rapidly after settlement, the excellent soils that Iowa started with were able to take a lot of mistreatment, so it took many decades for the effects of soil degradation to become obvious. While those effects have been clearly evident for some time, a growing number of environmental agencies and organizations are now monitoring and documenting soil degradation and making their conclusions readily available in an effort to inform and advise the public.

CHAPTER SIX

Squandering the Inheritance

The soil is deep, of exhaustless fertility.
—IOWA BOARD OF IMMIGRATION, 1870

WITHOUT QUESTION, Iowa has played a significant role in feeding the world's people and livestock through the decades. According to the U.S. Department of Agriculture's 2017 Census of Agriculture, Iowa farmers harvested more than 2.5 billion bushels of corn and half a billion bushels of soybeans in one year (National Agricultural Statistics Service 2019). But the cost to our land has been tremendous, as highly mechanized agriculture, industrial chemicals, and new markets like ethanol have changed the landscape of Iowa. Today, we lead the country in tons of corn harvested for grain, but regrettably we also lead the country in tons of precious soil leaving our fields, too much of which ends up leaving the state.

The problem of soil erosion in the United States and around the world was recognized long before the Green Revolution intensified agriculture abroad. In the late 1930s, the USDA asked W. C. Lowdermilk to conduct a survey of land use in other countries in the interest of shaping conservation policy at home. It was the end of a decade that had seen the worst soil erosion in the history of this country. The denudation of vegetation and topsoil by unimaginable wind erosion during the Dust Bowl was accompanied by tremendous rates of water erosion during storms on land that had been stripped bare of topsoil and roots. That same decade, the Iowa Agricultural Experiment Station published *Soil Erosion in Iowa* (Walker and Brown 1936). Its abysmal findings led P. E. Brown to warn that "if the soil is not conserved . . . if we persist in the 'sins of mismanagement' . . . we are facing the prospect of national ruin sooner or later" (1936).

Lowdermilk was the associate chief of the Soil Conservation Service. Renamed the Natural Resources Conservation Service in 1994, it was created in 1935 by the passage of the Soil Conservation Act after compelling testimony by Hugh Hammond Bennett before the Senate on the extent and consequences of soil erosion during the Dust Bowl years. As the story goes, Bennett made his case when a cloud of dust from the Great Plains more than a thousand miles away descended on Washington, D.C., during his testimony, darkening the sky and drifting in through the open windows.

The survey that Lowdermilk conducted of several nations in Western Europe, North Africa, and the Near East led him to conclude that agriculture and land abuse over hundreds and in some cases thousands of years had helped topple empires, while careful stewardship of the land by other societies had helped them flourish for centuries. His 1953 report is still very relevant and in continual demand despite the many changes in land management that have taken place since then. More recently, David Montgomery's book *Dirt: The Erosion of Civilizations* gives an excellent and detailed account of soil erosion and its consequences at home and abroad, both over the course of history and in the present day. My goal here is to shine a light on Iowa to show how our premier agricultural state fits into this global and national pattern of unsustainable treatment of soil and how the tremendous costs to our state—and our downstream neighbors—continue to accrue as a result.

Lowdermilk came to believe that the greatest problem through the ages had been trying to establish permanent agriculture on sloping land, a conclusion that applies especially well to Iowa given its hilly topography. After the establishment of the Soil Conservation Service, efforts to reduce Iowa's erosion rates began in the 1940s and 1950s with much government persuasion and assistance. Some of the earliest practices to be installed were terraces—earthen embankments—contour farming, and strip cropping on sloping ground, as well as windbreaks. Iowa has more miles of terrace than miles of stream, and installation of terraces continues to this day. Along with newer Natural Resources Conservation Service practices like riparian buffers along streambanks and grassed waterways, terraces trap eroded sediment and phosphorus and intercept or slow the flow of runoff on slopes.

Leading the Nation in Sheet and Rill Erosion

Despite the outstanding conservation efforts of the Natural Resources Conservation Service and the Iowa Division of Soil Conservation and their thousands of dedicated employees over several decades, the consequence of farming on slopes in Iowa becomes strikingly clear when you delve into the numbers. Every five years since 1982, the Natural Resources Conservation Service in cooperation with Iowa State University has been conducting a survey of land use and soil characteristics in the United States, called the National Resources Inventory. The latest numbers available are from 2017, and they show that the average rate of water erosion from cultivated cropland was still almost 6 tons of soil lost per acre. That is significantly lower than it was in 1982 by about 28 percent (Natural Resources Conservation Service 2020a), and the improvement is of course welcome. But it's not as encouraging as it might sound—for the simple reason that erosion rates in 1982 were so high that most of Iowa, along with other parts of the Corn Belt, showed up as a glaring bull's-eye on a map of erosion rates in the United States (fig. 40).

As the 2017 map in figure 41 shows, some thirty-five years later and despite significantly lower rates of erosion, much of Iowa is still an unsightly orange blotch, and a mind-boggling amount of soil still leaves our state's fields every year—roughly 145 million tons according to the latest inventory. To put that into perspective, it's enough once-fertile soil to fill several hundred thousand freight cars, which would form a train long enough to stretch from Council Bluffs to Davenport more than twenty-five times. Every year. Even more alarming is research out

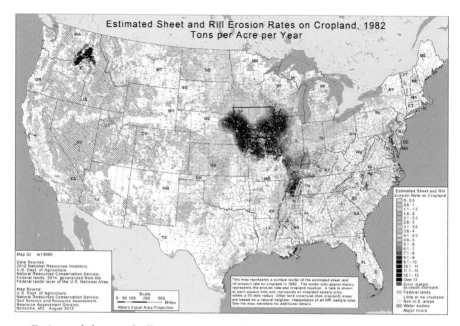

40. Estimated sheet and rill erosion rates on U.S. cropland in 1982. From an updated map in the 2012 National Resources Inventory.

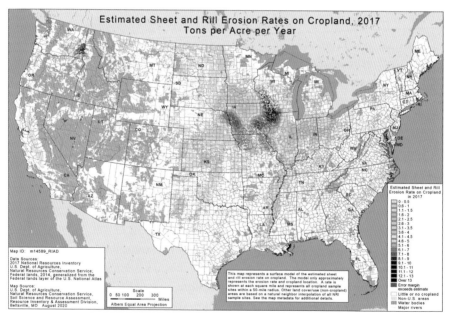

41. Estimated sheet and rill erosion rates on U.S. cropland in 2017. From the 2017 National Resources Inventory.

of Iowa State University that shows that this amount of erosion may represent only about *half* of what is really taking place on the land.

The National Resources Inventory does not provide numbers on how much of this sediment ends up in waterways or how much eventually leaves the state, although even a small percentage of those 145 million tons every year is a lot! But even within Iowa's borders, topsoil is displaced from where it is needed the most. It moves from fields on hills and slopes, which constitute the majority of agricultural land in Iowa, to lower positions on the landscape—footslopes, toeslopes, and floodplains. Not only does this result in acreage that is much less fertile for agriculture, these lower landscape positions are vulnerable to flooding and a high water table, conditions that hamper cultivation. Of course, far too much of the sediment does end up in our state's streams, lakes, water supply or recreational reservoirs, and flood-retarding ponds, where it becomes a water-quality problem and a financial burden—for example, see the discussion of lake sedimentation below. The Natural Resources Conservation Service's procedure for estimating ballpark percentages of the eroded soil that gets delivered to water bodies varies widely according to the size of the watershed and its location—from less than 10 percent in flat north-central Iowa to as much as 90 percent in the Loess Hills of western Iowa.

In order to comprehend what's happening on the ground, it's important to understand exactly what constitutes soil erosion in Iowa and how it is measured or estimated. The rates collected by the National Resources Inventory are twofold: first, soil erosion from water flowing over cropland, cropland converted to grassland, and pastureland; second, wind erosion. Wind erosion can be a serious problem in parts of Iowa, especially during the winter when fields lie bare and blackened snowdrifts can be seen along many of the state's highways. It must not be dismissed, but the state's average rate of wind erosion is much lower than it is in states like North Dakota, Kansas, Oklahoma, Texas, and New Mexico. However, when it comes to water erosion, Iowa is the indisputable king.

Over the years, the USDA's Agricultural Research Service has developed procedures for estimating soil loss due to sheet and rill erosion—the removal of soil from the land surface by the action of rainfall,

42. Rill erosion on cropland in western Iowa. Photograph by Gene Alexander, USDA–Natural Resources Conservation Service.

snowmelt, and runoff. The sheet portion refers to a fairly uniform layer of soil, while rills are small channels only a few inches in depth that tillage operations are able to obliterate each spring, only to have them form again over the course of the year (fig. 42).

To estimate rates of sheet and rill erosion, the Natural Resources Conservation Service used the Universal Soil Loss Equation beginning in the 1970s and the Revised Universal Soil Loss Equation beginning in the 1990s until, in 2019, it began using a different, more accurate model. All these models calculate soil erosion in tons per acre per year, using various equations that quantify the interacting factors controlling the rate of erosion—rainfall, soil erodibility, topography, vegetation, and land management. Farmers can even find a soil erosion calculator at Iowa State University's Extension website to estimate erosion on their own fields.

A ton of eroded soil has the volume of a cube about a yard long on each side, which may not seem like a lot, but when five or six times that much is applied to the 25 million acres of cultivated land in Iowa, it adds up to that freight train stretching across Iowa many times over. According to the National Resources Inventory, the average rate of sheet and rill erosion on cultivated cropland in Iowa in the early 1980s was a whopping 8.10 tons per acre per year, so our imaginary train of displaced soil was even longer back then.

By 1992, the average annual rate of erosion had dropped by 26 percent to 5.96, largely due to the passage of the Food Security Act of 1985, which produced the single greatest improvement in farming practices in decades. It included a farm bill provision requiring producers to get serious about conservation compliance in order to remain eligible for farm subsidies. By 1992, this had resulted in more than 2 million acres of cropland being taken out of production altogether and converted to noncultivated uses—native grasses, riparian buffers, plantings for wildlife habitat, and other uses—through the newly created Conservation Reserve Program. After 1995, this program targeted highly erodible land, which the USDA defines as land where the sheet and rill erosion rate on unprotected soil is at least *eight times* the supposed tolerable rate for that soil. It's almost unfathomable that such land was ever farmed with no protection in place—no terraces, contour farming, prairie strips, or like practices—but unfortunately, in too many cases, it still is.

Since 1992, the average rate of sheet and rill erosion as estimated by the National Resources Inventory has fluctuated only slightly. The latest numbers, from 2017, show the average rate of sheet and rill erosion on cultivated land in Iowa to be 5.79 tons per acre per year. At first glance, this may not seem a lot higher than the T value of 5 assigned to 70 percent of Iowa's soil series. The T value, which stands for "tolerable soil loss," is supposedly the maximum amount of soil that can be removed annually without degrading long-term productivity. The other 30 percent of Iowa's soil series have T values lower than 5, meaning that they are even more sensitive to soil loss.

Then again, if we recall the time factor of soil formation from chapter 3, is a T value of 5 even realistic? Many experts from a cross section of agricultural and environmental fields don't think it is. The T values currently in use were based on the best judgment of informed soil scientists several decades ago, rather than on rigorous research (Brady and Weil 2008). They were considered economically and technically feasible in the 1950s, but most agronomists today advocate for soil erosion of less than an inch in 250 years to maintain productivity. Others maintain that a soil replacement rate of an inch every 500 to 600 years is realistic for loess soils of the Great Plains (Logan 1982; Montgomery 2008). Iowa's "tolerable" rate of 5 tons per acre per year translates to an average of one inch of soil washing away in only about 40 years—six to twelve times greater than these suggested rates. With actual erosion rates often far exceeding the T value, it's not difficult to understand how Iowa fields have lost, on average, 6 to 8 inches of topsoil in the 180-plus years since the steel plow was introduced.

To make matters worse, the total area of cultivated cropland in the state has steadily increased since 1992 by nearly 5 percent, more than a million acres. As of 2017, there were 25 million acres under cultivation, which is the most of any state and the most since 1982, before passage of the 1985 farm bill provision in the Food Security Act. During those same twenty-five years, from 1992 to 2017, the amount of Iowa land in the Conservation Reserve Program decreased by an astonishing 67 percent to only 698,000 acres, with most of the decrease coming since 2000 and coinciding with the ethanol boom. The rental rate—usually less than $150 per acre—that the government was paying farmers for keeping their environmentally sensitive land in the program could not compete with the profit potential of growing corn for ethanol production. A spike in corn and soybean prices, which resulted from widespread drought in North America from 2010 to 2012, also led farmers to take acres out of the program. Then in 2014, a new farm bill lowered caps on the total number of acres the program could fund by 25 percent, and although the 2018 farm bill raised the cap somewhat, it is still 15 percent lower than it was in 2013.

The combined effect of high erosion rates and steadily increasing acreage under cultivation means that Iowa has maintained the dubious honor of leading the nation in tons of soil lost from cropland since the National Resources Inventory began publishing numbers in 1982. Despite having only 7 percent of the nation's cropland, Iowa produces 15 percent of the country's eroded soil from cropland. Illinois is second, with about 7 percent of U.S. cropland and about 10 percent of the eroded soil. Together, the states of Iowa, Illinois, Missouri, Nebraska, and Kansas are responsible for nearly half of the country's sheet and rill erosion on cultivated land.

As if these numbers weren't bleak enough, a national research consortium's recent investigation into erosion on Iowa cropland painted an even more discouraging picture. The Environmental Working Group's national scope of research is broad, including land, energy, food, water, and health. EWG Midwest, based in Ames, focuses on farming and natural resources policy. In 2011, EWG Midwest released an eye-opening report entitled "Losing Ground," which warned that "across wide swaths of Iowa and other Corn Belt states, the rich, dark soil that made this region the nation's breadbasket is being swept away at rates many times higher than official estimates" (Cox, Hug, and Bruzelius 2011). The report was front-page headline news in the *Des Moines Register* and made the *New York Times* a few weeks later in an editorial titled "Washing Away the Fields of Iowa."

The study drew upon data gathered daily since 2003 by a collaboration of Iowa State University and USDA scientists as part of the Iowa Daily Erosion Project. An enterprise possible only with today's advanced computer technology, the innovative project uses the USDA's latest soil erosion model, the most current databases of soil and land-use characteristics, and radar-derived weather data to calculate daily estimates of rainfall, runoff, and soil erosion for the state. The cutting-edge aspect of the project is its ability to generate data for agricultural land after every single storm that hits anywhere in the state. In contrast, estimates in the National Resources Inventory are based on rainfall amounts averaged over many years. The Daily Erosion Project calculates average daily erosion rates for each and every one of Iowa's 1,580 legal townships, which is an enormous improvement in precision over the national inventory's five-year statewide average. (Legal townships are surveyed squares of land 6 miles on each side, encompassing 23,040 acres.)

According to EWG Midwest's report, the daily erosion data for the sample year of 2007 showed nearly a quarter of Iowa's cropland, 6 million acres, eroding twice as fast as the supposedly tolerable rate. In some parts of the state, storms triggered soil erosion that was twelve times greater than the statewide average rate reported in the 2007 inventory.

The new information in "Losing Ground" did not invalidate the inventory's reported average erosion rate for Iowa cropland. The Daily Erosion Project reported a statewide average rate for 2007 that was *less* than the inventory's rate. What the report revealed is that average rates obscure what is really happening in large areas of the state, where the erosion occurring during some storms often exceeds the tolerable level by several times—for example, exceeding 50 tons per acre per year in eight townships in 2007. In fact, in early May of that year, a storm that ravaged southwest Iowa resulted in sheet and rill erosion rates on some unprotected fields of 100 tons per acre in a single day!

Adding Gully Erosion to the Equation

As a second component to their investigation, EWG Midwest corroborated the findings of the Iowa Daily Erosion Project for the year 2020 with aerial surveys. In the process, they saw striking visual evidence that another common form of erosion on Iowa cropland is far worse than previously thought. This type of erosion occurs by way of ephemeral gullies, which are shallow ditches that form in natural depressions where runoff from rainfall concentrates (fig. 43).

These gullies are called ephemeral because they can be reshaped, smoothed over with a disc, and planted across each spring. Farmers usually fill them in with loose erodible topsoil, often previously treated with agrochemicals, which gets washed away during rainstorms over the course of the year, over and over again, year after year. Estimates by the Natural Resources Conservation Service are that 50 to 90 percent of the eroded soil ends up in local creeks, and much of that is eventually transported to larger and larger streams in the watershed. If left unchecked, some turn into deep permanent gullies that work their way upslope into fields, beginning where the ephemeral gullies discharge into the creek. These "classic" gullies, which can easily grow to 10 feet deep and as much as 40 or 50 feet deep in western Iowa, end

43. Roadside view of sediment at the end of ephemeral gullies in central Iowa, consisting of eroded topsoil. Photograph by Lynn Betts, USDA–Natural Resources Conservation Service.

up delivering tremendous amounts of sediment and other pollutants to Iowa's streams and lakes.

Neither the National Resources Inventory nor the Iowa Daily Erosion Project provides estimates of soil loss from ephemeral gully erosion. Gully erosion on the landscape is not included in the USDA erosion models and is notoriously difficult to estimate. Procedures exist to measure the volume of soil lost from an individual gully, but determining the number of ephemeral gullies that appear over large areas during the course of the year is much too time-consuming on the ground to be feasible, even before the gullies become concealed by a canopy of corn. EWG Midwest's aerial surveys, conducted after spring rains when ephemeral gullies are most visible, were able to document the extensive damage seen from the air in videos made available to the public. The flights found numerous fields scarred by gullies, which runoff had rapidly carved in a matter of one or two days. Based on the surveys and other published studies, EWG Midwest's report concluded that if the National Resources Inventory included ephemeral gully erosion in its estimates, reported soil loss could more than double.

Two years after releasing "Losing Ground," the group followed up with "Washout," which provided an interactive map for online readers showing general locations and photos of some of the devastating gully erosion that occurred on highly erodible land in late May 2013, the wettest spring on record at the time (Cox, Lorenzen, and Rundquist 2013). It presented Iowa Daily Erosion Project estimates indicating that, in five days alone, 1.2 million acres of Iowa cropland lost more soil from sheet and rill erosion than the 5 tons per acre considered tolerable for an entire year. Some 4,000 acres of poorly protected farmland— farmland lacking conservation practices—lost more than 20 tons per acre in that one week. "Washout" quoted an editorial in the *Storm Lake Times* saying, "Watching the inlet to Storm Lake on Monday, it looked like Willy Wonka had just opened the chocolate milk spigot. Upstream the erosion was sickening." Viewing many of the photos on EWG Midwest's interactive map of just nine central Iowa counties elicits the same reaction. In my own experience evaluating erosion in Iowa, I was appalled too often by the sight of raw gullies on cropland, the ugly tons of black dirt piled thick on field bottomlands, and the disconcerting smell of fresh mud, very distinct from the aroma of fertile living soil.

The same map also takes viewers to photos of fields where conservation practices were working to protect the soil. This observation was reinforced the following summer, when the group's researchers revisited eighteen of the crop fields they had surveyed and photographed in 2013. They found some better news, mainly due to the fact that the nine counties were not hit as hard by storms that spring. In addition, however, farmers had implemented new conservation measures on five of the eighteen tracts and were maintaining soil-saving practices already in place on another five tracts. Unfortunately, eight fields showed no signs of enhanced soil conservation measures at all (Rundquist and Cox 2014).

EWG Midwest's unrelenting message has been that too many producers are out of compliance with the practices in their conservation

44. Grassed waterway for preventing ephemeral gully erosion in a Fayette County cornfield. Photograph by Tim McCabe, USDA–Natural Resources Conservation Service.

plans for which they are receiving federal payments, and we need to get back to full enforcement of conservation compliance. In 2017, the Natural Resources Conservation Service issued a national flyer titled "Fix It, Don't Disc It," recommending several management options and conservation practices available to help farmers voluntarily fix ephemeral gullies in their fields. In Iowa, the agency notified farmers participating in USDA programs that they would "now be required to provide additional control of ephemeral gully erosion on their highly erodible fields." Whether this was related to EWG Midwest's reports is unclear. Officially, it was in response to recommendations from the Office of Inspector General for the agency to modify its compliance review procedures.

The agency had been funding the installation of grassed waterways, one of the main practices for preventing the formation of ephemeral gullies, for many years. Grassed waterways are broad, shallow channels engineered to the proper shape and width and seeded with a vegetative cover (fig. 44). When properly maintained, they slow the velocity of runoff and convey it to a stable outlet where it enters a natural waterway.

Between 2013 and 2018, the Natural Resources Conservation Service earmarked $2.8 million toward the practice, which funded at least 850 grassed waterways in sixty-six counties (Mustapha Abouali, personal communication). The Iowa Division of Soil Conservation has also provided cost-share money for hundreds of grassed waterways across the state, as well as a few thousand terraces and other practices that prevent ephemeral gully erosion. In 2019, the USDA selected Iowa as one of six states to take part in a pilot program that provides $5 million of additional financial assistance to Iowa farmers to address ephemeral gully erosion on highly erodible land. All this represents a good start, but the scope of the problem is enormous. A recent evaluation of aerial imagery by EWG Midwest found 119 miles of eroding flow paths in a 66,000-acre sample of Iowa's highly erodible cropland (Cox and Rundquist 2018). Extrapolating from that sample, there could be more than 15,000 miles of eroding flow paths in Iowa's 8.5 million acres of highly erodible land. Judging from my twenty years of personal experience evaluating gully erosion in Iowa, I can easily believe that there are at least 75,000 ephemeral gullies scarring Iowa fields each spring.

As we'll see in the following chapter, there are much better ways to address both sheet and rill erosion and gully erosion than installing terraces and grassed waterways to simply manage the runoff and collect the sediment. There are innovative as well as tried-and-true practices that reduce the amount of runoff in the first place by increasing infiltration of water into the soil, right where the rain falls. And they are lasting solutions rather than Band-Aids with limited life spans like terraces and waterways, which catch the blood but don't stop the bleeding.

The costs of soil erosion are enormous, both in terms of money and in terms of damage to our natural resources. The least visible and yet the most widespread cost is the damage inflicted on the soil itself. Obviously, the layer most affected is the topsoil—the most fertile layer, rich in organic carbon and other plant nutrients. So loss of topsoil is directly proportional to loss of fertility and lowered yields. In 2014, the

Des Moines Register reported on a study designed to determine the financial cost of this loss to the state, which led researchers to conclude that $1 billion in lost revenue each year from lowered crop yields alone was a very conservative estimate (Eller 2014). A couple of years later, researchers showed that the difference in dry corn yield between a 7-inch A horizon and a 15-inch A horizon is about 20 bushels per acre. Assuming a $4 price per bushel for corn, that would translate to a loss of at least $50,000 a year for a 640-acre farm (Cruse 2016; Cruse, Lee, and Sklenar 2016). None of these costs includes those associated with cleaning up drinking water, removing sediment from reservoirs and lakes, and numerous other direct and indirect impacts. An Iowa State University Department of Economics study estimated those costs nationally to be $12 to $39 per cropland acre per year (Tegtmeier and Duffy 2004).

Sedimentation in Iowa's flood-control reservoirs and recreational lakes is perhaps the most visible damage from erosion. One prominent example is Lake Red Rock in Marion County, a huge Army Corps of Engineers dam and reservoir where the loss of space for water storage is approaching 50 percent due to sedimentation. If you were to average out the amount of sediment reaching the lake over 365 days, it would be enough to fill seven Olympic-sized swimming pools each and every day (Thompson 2014b). Another example is Lake Panorama in Guthrie County, the largest private lake in Iowa. Dredging in different parts of the lake goes on year round, when it's not covered with ice, at the cost of at least $2 million a year. With nearly three-quarters of the lake's Middle Raccoon River watershed in row crops, the river delivers more sediment from soil and streambank erosion than dredging can keep up with. This despite the removal by dredging of up to 400,000 cubic yards of sediment each year, which is enough to fill nineteen football fields to a depth of 10 feet! Where to continue dumping all this dredged sediment is a looming problem. However, despite the financial cost and the challenges, the lake sponsors have sadly realized that getting landowners in the watershed to decrease soil erosion is actually more challenging (Thompson 2017). It's even more problematic because many of the landowners rent their land to tenant farmers, who may farm it only for a few years and who do not have long-term invested interest in the health of its soil.

Much of the sediment ending up in these lakes and reservoirs comes from another source in addition to eroding fields and gullies. During heavy rain events, gullies quickly deliver massive amounts of runoff to streams and increase the amount of streamflow many times over. This deep and rapid streamflow attacks streambeds and streambanks, causing channels to deepen and widen and tremendous amounts of sediment to travel downstream. In many if not most cases around the state, these banks and floodplains are themselves made up of sediment produced by past erosion of fertile soils in the adjacent fields.

What Tillage Does to Soils

Even on flat landscapes where soil erosion is minimal, soil fertility has suffered greatly, primarily from intensive tillage. Although more and more of the state's farmers have used alternative conservation tillage practices for the past thirty to forty years, in 2017 nearly half of Iowa's cropland acres were still being tilled by conventional methods, according to the USDA Census of Agriculture. That generally means using a chisel plow, which buries up to 70 percent of the previous year's crop residues and leaves the rest on the surface, but the moldboard plow is still used to actually invert the soil on some acres in Iowa. Plowing may be followed by disking to cut up the residues and incorporate them into the soil. Next, one to three passes with a harrow breaks up the large clods of topsoil to create the seedbed.

All these tillage activities expose more and more soil material to the air. When soil carbon—perhaps thousands of years old—is exposed to the air, it combines with oxygen to form carbon dioxide, which is then lost into the atmosphere. (From this, you might correctly deduce that what decreases soil fertility also increases the carbon dioxide content of the atmosphere, the main culprit in our current rise in global temperature.) Exposure to the air also increases the metabolism of the soil's organic matter by microbes, which I will discuss in greater detail in chapter 7. The practice of removing and baling plant residues after harvest adds another blow to soil fertility, as it steals food from soil macroorganisms like earthworms. Soil life simply hasn't evolved to cope with the removal of dead plant material on a regular basis, which just doesn't happen in nature.

Furthermore, when tillage exposes soil peds to the surface, especially in the absence of crop residues or plant canopy, they become extremely vulnerable to the beating action of raindrops, which can hit the ground at speeds up to 20 miles per hour. When rainfall is more intense—that is, a lot of rain in a short period of time—the drops are larger and fall even faster. The dispersed soil particles settle into a thin dense layer that forms a hard crust when it dries. The crust can delay or prevent seedlings from emerging and cause ponding of water. Most importantly for environmental quality, the greatly reduced infiltration leads to high rates of runoff, which will only worsen as intense rain events become more and more common under a changing climate.

The destruction of soil structure from tillage results in compaction of the plow layer itself, and compaction is a very significant issue in our state. An Iowa Farm and Rural Life poll showed that at least 75 percent of Iowa farmers were concerned about compaction of their soils (Arbuckle and Lasley 2013). If the soil is tilled when it's too wet, the peds basically get churned into mud that has lost all soil structure and has little to no porosity. When this structureless mass dries it becomes hard, leading to poor root penetration and reduced infiltration of rainfall. Destruction of porosity also leads to poor aeration of plant roots, which get most of their oxygen through macropores, those pore spaces wider than about a tenth of a millimeter. So the right proportion of air and water is vital for a healthy root system. Unfortunately, with Iowa's spring rainfall on the upswing in recent years, wet soils are becoming unavoidable and conventional tillage more and more damaging.

After the tilling is done and the crop is planted, farmers may make several passes with an herbicide sprayer or a row cultivator to kill weeds. Finally, tractors pulling large combines and grain carts show up for the harvest. In other words, heavy equipment passes over the field several times each growing season, compacting the soil, crushing the peds, and obliterating the soil pores—or what's left of them

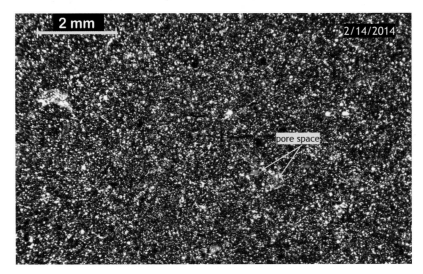

45. Compacted topsoil with very low porosity from a conventionally tilled field. Compare these small pores (blue) to the large pores (white) in figure 7 in chapter 1, taken at the same magnification. Photograph by the author.

(fig. 45). The size and weight of farm equipment have increased as farm size has grown, to the point where the soils in our crop fields experience heavier loads than most of our roads, which were engineered for that purpose. Most public right-of-way regulations require axle loads of less than 10 tons, but many large combines and grain carts exceed that by 50 percent (Hanna and Al-Kaisi 2009). When full, some grain carts can have axle loads greater than 38 tons.

Compaction isn't limited to the plow layer or to the topsoil. Over the years, the combination of heavy farm machinery and other downward pressure from tillage has created a very compacted layer in many of our field soils at a depth of 6 to 7 inches, immediately below the plow layer and often below the topsoil. This layer, known as a plow pan, impedes root penetration, especially when it dries out and hardens. A plow pan easily exceeds most crop roots' level of resistance to penetration, which is about 300 pounds per square inch (Magdoff and Van Es 2009). Instead, roots grow short and thick above the plow pan, with few

46. Rainwater in northeast Iowa infiltrates rapidly into the no-till field on the left. In the conventionally tilled field on the right, much of the water runs off and eventually reaches a nearby creek, likely carrying farm chemicals with it. Photograph by Neil Sass.

fine roots and root hairs to gather nutrients, and are extremely vulnerable to dry weather as a result. Conversely, water can perch above the plow pan and cause waterlogging of the topsoil during wet weather.

When you drive by fields after a rainstorm and see standing water, this is often a sign of soil compaction or surface crusting or both. On sloping fields where ponding is less likely, water unable to infiltrate into the soil concentrates instead and forms runoff, leading to sheet and rill erosion and the formation of ephemeral gullies. Runoff water from Iowa's agricultural fields transports sediment, phosphorus, nitrogen, and other pollutants to streams and lakes. According to a University of Iowa study reported in the *Des Moines Register*, Iowa contributes a shockingly large share of the nitrogen going into the Upper Mississippi River (45 percent) and the Missouri River (55 percent), despite comprising a much smaller portion of both watersheds than other states (Jones et al. 2018). For instance, Iowa makes up only 3.3 percent of the Missouri River Basin. The same study estimated that 90 percent of the nitrogen contaminants in Iowa waters can be sourced to land under cultivation, with point sources such as industry, concentrated animal feeding operations, animal waste facilities, and wastewater treatment plants supplying the rest. Although many Iowa farmers currently practice no-till cultivation, far more do not, as will be discussed in chapter 8. Foregoing tillage before planting makes an enormous difference in how much rainwater can infiltrate the ground instead of running off the field (fig. 46).

Conservation efforts since the 1940s have prevented significant amounts of runoff and sediment from leaving the farm, but they clearly have not been even close to adequate. The enormous and widespread consequences of runoff in Iowa cannot be solved without addressing the underlying cause of the runoff, which is poor infiltration of water into our soils. In the words of Ray "The Soil Guy" Archuleta, a former soil health guru for the Natural Resources Conservation Service, "We don't have a runoff problem, we have an infiltration problem." Conservation tillage goes a long way toward improving infiltration, but increasing soil organic matter is also critically important. For any given soil texture, an increase in organic matter improves infiltration and significantly increases the amount of water a soil can hold on to and keep available for plant use (Hudson 1994). Of course, it's a cyclical process, because the growth and decomposition of plants are necessary to maintain that organic matter. If we are to abandon harmful tillage practices and reduce or eliminate the inorganic nitrogen that is impairing our waters, the soil system will have to become much more dependent on the life within it—its microbes, insects, and other fauna. As we'll see next, they are the only agents capable of keeping the soil aerated and converting many of its necessary nutrients into forms usable by plants.

CHAPTER SEVEN

Rediscovering the Living Soil

*I had my cameraman set up a shot focused on a patch of soil.
I called out, "Action!" He turned to me and said, "There's nothing
happening." "Oh," I replied, "there is, there is! There's so much
happening! How will we fit it all in?"*

—DEBORAH KOONS GARCIA, 2014,

ON FILMING *SYMPHONY OF THE SOIL*

PART 1 OF THIS BOOK covered the physical and chemical properties
of soil, with a promise that the biological properties of soil would have
their own chapter. As you will soon see, living organisms are in essence the principal drivers of soil fertility and long-lasting soil health.
Unfortunately, this is a fairly recent discovery—or rediscovery—which
has come about in reaction to the looming soil crisis resulting from
the mismanagement described in chapter 6. Aside from organisms—
principally vegetation—as one of the five soil-forming factors, the science of soil traditionally gave only cursory attention to soil ecology,
generally emphasizing only the influence of macroscopic animals on
soil aeration and certain forms of soil structure. I recently looked over
my notes from a soil fertility class I took in 1988 and could not find a
single note about earthworms or soil biology in general. It may have
been an omission on my part, because the 500-page textbook for the
course that still sits on my shelf does contain 6 pages of text about living organisms in the soil. On the other hand, it also contains 160 pages

addressing the amendment of soil with fertilizers and the statement that "excessive use of fertilizer is still rare even in the United States . . . [and] there is much room for increased fertilizer usage." It's not that difficult to see how we got to where we are today.

In the latter half of the twentieth century, down on the farm where the science got applied, fertile soil came to simply mean soil with adequate drainage that had the appropriate balance of nutrients to produce a successful crop, however that balance could be achieved. In a matter of a few decades, applying expensive fertilizers—some of which are toxic to soil life—became the method of choice for providing crops with the nutrients they needed. That pattern has persisted into this century. As an example, in the 2018 crop year alone, farmers applied 3.8 billion pounds—1.9 million tons—of chemical fertilizers and 55 million pounds of pesticides to Iowa's corn and soybean crops (National Agricultural Statistics Service 2019).

This reliance on chemicals has led us to unsustainable industrial-scale agriculture. It represents our utter failure to think ecologically and is just one part of a disconnect with nature that characterizes modern life in general. While the agricultural chemical industry has thrived many times over, with more than $41 billion now being spent every year on farm chemicals in the United States, our soils are suffering a serious decline in the number and diversity of soil organisms. We are effectively ignoring and destroying the biology of our soil in this country.

Such disregard for the soil didn't always exist, of course. Throughout history, humans from many traditions considered soil to be alive and sacred. The creation myths of many Native American cultures invoked life-forms sculpted from the soil, and reverence for the soil was part of their worldview. Likewise, indigenous peoples in Australia believe that the soil is filled with divine presence and holds the life forces of all the species on Earth. In the Judeo-Christian tradition, the biblical story of Adam and Eve honors the marriage of soil and life, for the Hebrew word *adamah* means "earth" or "soil" and the name Eve comes from the Hebrew word for living, *chavah*. There are many other examples.

Even before the Green Revolution with its incredible bounty, not everyone bought into the seemingly miraculous solution of manufactured chemicals. In the 1930s, a couple of decades after the invention of synthetic ammonia, the English botanist and agronomist Albert Howard began talking about a Law of Return—the necessity of returning organic matter to the soil in order to produce healthy crops free of the fungal diseases that seemed to accompany the use of chemical fertilizers. Eve Balfour, an early pioneer of the organic farming movement in the 1940s, stressed the need to adopt farming techniques that maintained soil fertility indefinitely by imitating natural systems and relying on the regenerative resource of living organisms. Other early critics of the sterile concept of soil and farming, such as Masanobu Fukuoka in Japan and Jerome Rodale in the United States, stressed the fallacy of relying on chemical fertilizers to the detriment of living organisms and organic matter in the soil. For the most part, their preaching fell on deaf ears, except for the subset of innovative growers who began to practice organic farming, which by its definition requires a reliance on life in the soil to maintain fertility.

As of 2017, there were about 780 certified organic farms in Iowa with several more awaiting certification, out of a total of 86,000 farms in the state. There is much skepticism in the agricultural community about whether organic farming can compete with or succeed on a scale as large as Iowa's millions of acres of corn and beans, despite examples where it has (see chapter 8). But whether you believe USDA-certified organic farming is feasible for large-scale agriculture in Iowa isn't the real issue. Sustainable farming may not require a choice between two extreme alternatives—strictly regulated organic farming or status quo chemical-heavy farming. In a way, this current dichotomy of thinking may be self-defeating. Other terminology used around the world, such as biological, ecological, regenerative, or low-input farming, may be more instructive. Any system that drastically reduces off-farm inputs and maintains soil fertility by letting soil organisms produce the nutrients—and earthworms do the plowing—can be considered good farming. There are many small farming operations as well as larger farms in Iowa that are not certified organic but that use no chemicals or very few chemicals and only in special circumstances. These farmers choose instead to rely on the biology of their soils. They are practicing what David Montgomery calls organic-ish farming (2017).

Regardless of what this new way of farming is called, since the 1990s we have witnessed a growing emphasis on sustainable agriculture around the world, and the movement has begun to put down some deep roots in our state. The recognition of a growing soil crisis and the establishment of sustainable agriculture as a major objective by the United Nations Earth Summit in 1992 may have heralded the beginning of this movement. By 2004, it had gained enough steam to earn a special issue of the premier scientific journal *Science*, which reported that "interest in soil is booming" due to technical advances in microbiology and microscopy for the study of the minute soil organisms called microbes (Pennisi 2004). The United Nations declared 2015 the International Year of Soils, which spurred numerous educational events, symposia, and online activities around the globe, and the International Union of Soil Sciences followed up by proclaiming 2015 to 2024 the International Decade of Soil. An award-winning feature-length documentary called *Symphony of the Soil* filmed on four continents examined our human relationship with soil, "this miraculous substance," and its complex and dynamic nature (Garcia 2012). Books about soil biological health and regenerative farming and gardening have sprung up like spring wildflowers, including *Teaming with Microbes*, *Life in the Soil*, and *Growing a Revolution*, to name just a few. We now recognize December 5 as World Soil Day, and simulated and interactive exhibits featuring living soil have appeared in numerous museums around the world. The Smithsonian's *Dig It! Secrets of the Soil* exhibit now has a permanent home at the St. Louis Science Center after traveling to several U.S. cities since it was launched in 2008.

Central to all these testimonies and activities has been an emphasis on the biological health of the soil or, in short, soil health. In this revitalized twenty-first-century way of thinking, the physical structure—the architecture—and the chemistry of soil are viewed as the stage upon which soil organisms act out their life stories, enriching and regenerating the soil in the process. We no longer think of soil as simply a medium consisting of mineral and organic matter that exists for the purpose of growing plants. It is a living ecosystem that demands our respect and attention just as much as the tropical rainforest or coastal waters. Respecting the soil requires us to recognize it as a living substance, a complex organism that is currently suffering from decades of abuse and neglect (Gershuny and Smillie 1999). With this in mind, it might be fitting to ask yourself, What soils live in my garden or my fields?

It has become abundantly clear to most scientists involved with soils, agriculture, horticulture, or the food supply that our current reliance on chemicals and widespread abuse of the soil is unsustainable. It will fail to meet the challenge of feeding the burgeoning population of the future and could result in a large-scale disaster instead. We need a new revolution to repair the damage that resulted as an unanticipated but tragic side effect of the Green Revolution. The growing soil health movement must become the agricultural revolution of the twenty-first century, in tandem with sustainable management of our other natural resources, if we are ever to rescue the health of the planet and its inhabitants.

And there is hope. Chapter 8 describes the exciting changes in agricultural practices taking place in Iowa as part of this national and international soil health movement, and it highlights some of the groups and farmers who are leading the charge. First, however, to lay the groundwork for understanding what we need to do to transform farming into an ecological and sustainable endeavor, in this chapter we will take a look at soil as a living ecosystem. I will explain the fundamental importance of soil organic matter—both living and dead—to the health and fertility of the soil, describe the rich diversity of life found in the soil, and discuss what soil organisms do to keep soil healthy. This will help you understand what one Grinnell farmer means when he says that, rather than feeding the plants with fertilizers, it is essential to "feed the *soil* and give it a roof over its head."

A New Direction at the USDA

Coming from a geology background, prior to about 2010 I was just as guilty as most in my view of soil as basically a nonliving physical substance. From my excellent graduate course in soil genesis at Iowa

State University, which included some soil chemistry and physics, and another class in soil fertility, I knew something about soil organic matter and the nutrients plants need. But in 2008, I attended my first soil health training session. It was the beginning of the Natural Resources Conservation Service's new soil health initiative, which started during the tenure of U.S. Secretary of Agriculture Tom Vilsack, former Iowa governor and environmental advocate. That's the day it dawned on me. Those little bugs in my backyard prairie plot, both the ones I could see and the microscopic ones I washed off my hands before eating lunch? They were playing a whopping role in making my soil fertile. And they could be doing it all over Iowa's fields if we would only quit killing them and give them half a chance!

This was new and exciting stuff, and it wasn't long before I saw it as an opportunity to expand my role in the agency and work with soil in another capacity. Since I had previous experience with a microscope from my graduate studies, our state soil scientist, Rick Bednarek, decided to trust me with his newly purchased microscope to look for and photograph microbes in soil samples. Of course, my experience with microscopes was limited to a geology class in grad school looking at ancient pollen and seeds from bogs, as well as extensive use of a petrographic microscope to look at resin-infused thin sections of soil. In other words, I was used to looking at dead things, not tiny critters that danced and zoomed around under the light of the scope! But I bravely ordered a few supplies, set up a simple soil health lab, and started looking. With a lot of patience and help from soil microbiologists like Tom Loynachan and Omar de-Kok Mercado of Iowa State, Elaine Ingham of the Rodale Institute, and Zach Wright, skilled brewer of compost tea in Fairfield at the time, I gradually began to understand what I was looking at. Rick provided me with soil samples that our soil scientists had collected from interested farmers wanting to learn about the life in their soils. In return, we sent them photos and short video clips of their farms' microscopic inhabitants and shared our results with field office staff and their farmer clients around the state, which piqued a surprising amount of interest.

Eventually, with retirement on the horizon, I handed over the microscope and the job to a couple of younger employees in the state who had some actual background in soil microbiology to go along with a ton of enthusiasm. In the meantime, the agency in Iowa started requiring soil health training for all its employees. This met with a lot of interest and helped raise the consciousness of many of my colleagues. Since 2012, the agency has had an excellent web page devoted to soil health. Despite the USDA's policy shift in 2017 to a renewed emphasis on production, which inevitably discourages conservation, the soil health genie was out of the bottle and seems to have survived the transition. The office of the U.S central region's soil health specialist is located in Des Moines, and the agency now has a soil health specialist on staff in most states.

So What Makes Soils Fertile?

In my soil fertility class back in the late 1980s, we learned the mnemonic phrase "C HOPKiNS CaFe is Mighty good, so CuMn Nick Zelany, before the MoB Closes it" for remembering the essential elements of plant nutrition—carbon, hydrogen, oxygen, phosphorus, potassium, nitrogen, sulfur, calcium, iron, magnesium, copper, manganese, nickel, zinc, molybdenum, boron, and chlorine. (In case you don't remember from chemistry class, the corresponding chemical symbols are C, H, O, P, K, N, S, Ca, Fe, Mg, Cu, Mn, Ni, Zn, Mo, B, and Cl.) The phrase was shorter when it was first coined right after Cyril Hopkins wrote his classic 1910 textbook on soil fertility, but it morphed over the decades as scientists recognized more and more plant nutrients.

   With the exception of iron, the first ten of these elements are macronutrients; the rest are considered micronutrients that plants need only in small amounts. Plants obtain carbon and oxygen from the air and oxygen and hydrogen from water in the soil, but the rest has to come from the soil solids. So for low-input sustainable farming, where these nutrients aren't added in the form of fertilizers, where will they come from? Well, most of them are already present in Iowa soils—some in the organic matter in what's left of the upper part of the soil profile, the A horizon—and others in the minerals contained within the soil's geologic parent material. However, most of the nutrients in soil organic matter are part of large and complex organic molecules that plants are unable to use. So here's the catch: living soil organisms must decompose the dead organic matter first, a process that converts nutrients into inorganic forms easily taken up by plant roots. For example, during decomposition proteins are converted into forms of nitrogen that are available to plants, namely, ammonium. Other microorganisms then convert much of the ammonium to nitrate, which is the main source of nitrogen for plants. Plants also get phosphorus and sulfur and most of the micronutrients they need from this conversion of organic molecules into inorganic mineral nutrients, all by way of living organisms as they feed to survive.

Other macronutrients such as calcium, potassium, and magnesium are added to the soil by the recycling of organic matter, but most come from the slow weathering of minerals contained in the soil parent material—in other words, the clay, silt, and sand particles. In Iowa, the sources for these nutrients are primarily the minerals calcite, dolomite, various feldspars, hornblende, pyroxene, and mica, all of which occur in either the glacial sediments or the bedrock or both. The calcium, potassium, magnesium, iron, and some other nutrients from these minerals occur as positively charged cations, dissolved in the soil water and available for uptake by plants. The problem is, they are also quite mobile in solution and leach into the subsoil every time sufficient rainwater moves through the topsoil. They remain available in the soil profile only if they can bond to negatively charged particles like organic matter or clay. A few of the micronutrients derived from soil minerals are kept available for plants by bonding tightly to small organic particles called chelates, by-products of decomposition that protect these trace elements from being converted to unavailable forms. The ability of organic matter and clay to hold on to nutrient cations is called the soil's cation exchange capacity, which is the amount of negative charge in the soil. It's a property that allows the gradual release of nutrients into the soil solution for use by plants during the growing season, so the higher a soil's cation exchange capacity, the better (Brady and Weil 2008).

This gets a little complicated, but I hope you can see that organic matter is of the upmost importance to soil fertility. The amount of organic matter in a soil is the most important factor affecting plant productivity, whether that is corn or big bluestem or tomatoes. For corn, a 1 percent increase in organic matter can increase yields by as much as 12 percent (Magdoff and Van Es 2009). This carbon-rich substance has other significant roles that contribute to soil health. It is time now to examine this substance—lumped as soil organic matter—more closely to see what it's really made of and what other functions it performs in a healthy soil.

In the main, soil organic matter consists of three things: living organisms (the living), fresh and decomposing organic residues (the dead), and stable, well-decomposed organic material (the very dead). Beside these three solids, biochemical compounds like amino acids, proteins, and root exudates exist in solution in the soil water. Root exudates are carbohydrates and proteins secreted by plant roots, analogous to the perspiration secreted through our skin.

The dead part of organic matter is the actively decomposing fraction, and it is the main food supply for the living organisms, which release the plant macronutrients when they feed. It also includes surface residues that protect soil peds from shattering under the intense impact of raindrops, which can throw soil particles as high as 2 feet into the air. This active fraction of organic matter is the part that responds most quickly to management changes. A significant decrease in this portion has a pronounced negative effect on the organisms that feed on it and on soil structure and porosity. As soil organic matter is depleted, problems with fertility, drainage, compaction, erosion, diseases, and pests become more and more common.

The organic matter I'm calling very dead is the fraction usually referred to as humus. There isn't a clear dividing line between decomposing and well-decomposed organic matter, as these two forms exist on a continuum rather than as two discrete classes. However, humus is typically older than a thousand years and is the fraction that is best at holding on to those positively charged cations of calcium, magnesium, iron, and others for gradual use by plants. Like clay, humus consists of very tiny particles, which means that it contains a lot of surface area. This allows for abundant chemical reactions such as attracting nutrients.

In addition to organic matter's role in maintaining soil fertility, sticky substances from the decomposition of plant residues help bind soil particles into the peds that make up soil structure. As a result, an increase in organic matter usually leads to an increase in the porosity of the soil and thus improves the infiltration of water. Not only is more water able to enter the soil profile, but organic matter, especially humus, is very good at holding on to water (Al-Kaisi and Kwaw-Mensah 2016). Porosity is the most important factor for water retention, but soil organic matter also has a significant influence. For example, one study found that an increase in organic matter from 1 percent to 4 percent doubled the amount of water readily available to plants, regardless of soil texture. This may be due to the fact that organic matter constituting 4 percent of a soil by weight actually makes up 15 percent of that soil's volume, to give just one example (Hudson 1994).

Estimates are that prior to the 1830s, Iowa's tallgrass prairie soils were able to absorb 90 percent of the rain that fell on the land through their extensive network of interconnected pores and abundant organic matter (Müller 2012). Before the prairies were plowed under, runoff happened only in the winter when snowmelt or rain flowed over frozen ground. Think of what that meant historically for water quality in Iowa. As I write this, at a time when most of Iowa's cultivated soils have both low organic matter and low porosity, more than 750 segments of Iowa's water bodies are impaired due to polluted runoff from farm fields, feedlots, and urban streets and parking lots (Iowa Department of Natural Resources 2018).

In the ideal soil, about 50 percent of its volume consists of pore space, with half of that space filled with air and the other half filled with water. Way back in 1943, Eve Balfour emphasized that it was impossible to overstate the importance of good soil structure and porosity, because together they ensure aeration, excellent water-holding capacity, and free drainage of the soil. If roots are able to penetrate deep enough as a result of strong structure, plants are able to find more water and nutrients by increasing the volume of soil they can exploit.

A perfect example of where deep-rooting vegetation and excellent soil structure work together is an Iowa prairie. The root system of a prairie plant might make up 70 percent of its weight and could be hundreds of years old. One square yard of big bluestem sod could have more than 25 miles of fine root hairs and rootlets (Müller 2012)! Some of those rootlets die each year, adding as much as 900 pounds of organic matter to each acre of prairie. Furthermore, the dramatic improvement of soil structure and porosity under prairie vegetation, which results in more water infiltrating the ground, also greatly reduces runoff, erosion, and the delivery of nutrient pollutants to streams. The conservation practice of prairie strips illustrates this benefit well. For instance, an experimental study concluded that strategically planting as little as 10 percent of a row crop field in prairie strips can reduce soil movement by 95 percent and phosphorus and nitrogen runoff by 80 to 90 percent (Schulte et al. 2017).

The Essential Life of the Soil

In addition to the deeper root penetration made possible by good soil tilth, plants have other ways of increasing the volume of soil from which they harvest water and nutrients. This is where living organisms come in. The living portion of soil organic matter includes roots and root hairs, many thousands of species of microscopic organisms grouped under the general term "microbes," and an astonishing variety of soil insects, earthworms, invertebrates, and vertebrates, including small reptiles and mammals. There may be as many as 100,000 types of organisms living in the soil, which exceeds the diversity aboveground by as much as a hundred times (Magdoff and Van Es 2009). It is this living community interacting with its nonliving environment that constitutes the soil ecosystem. In his delightful 2012 pamphlet, Iowa prairie restorationist and illustrator Mark Müller contemplates the dynamic life below the surface, where "predators and prey, life and death, sex, chemistry, deceit, disease, cooperation, warfare and magic exist right below your feet."

As with all life, the main goal of soil organisms is to survive—in other words, to find food, water, and air and to reproduce. Just like the

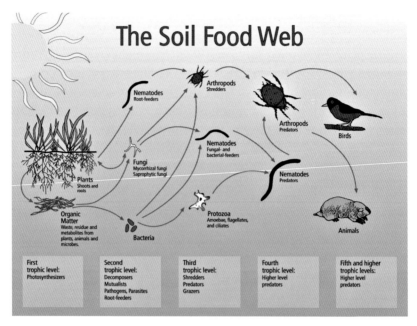

47. One depiction of a soil food web. From Elaine R. Ingham, Andrew R. Moldenke, and Clive A. Edwards, *Soil Biology Primer*. Courtesy of the Soil and Water Conservation Society and the USDA–Natural Resources Conservation Service.

food chain that exists in the living world above the surface, there is an order in which different levels of soil organisms feed upon one another and, in the process, convert energy and nutrients from one form to another. Not so much a chain as a soil food web, it's a network in which energy and nutrients are exchanged in a complex pathway that does not follow a straight line (fig. 47).

At the far left of the pathway, fueling the food web are the primary producers—photosynthesizers that fix carbon dioxide using energy from the sun to produce carbohydrates and proteins. They include plants as shown but also lichens, mosses, algae, and certain bacteria. At the next "level" are the primary consumers—fungi and nearly all bacteria, which feed on and break down or decompose the producers and their exudates. Bacteria are present in staggering numbers in the topsoil. For some perspective, there can be up to a billion bacteria in

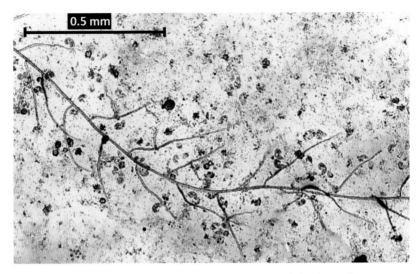

48. Actinomycetes, the branching bacteria. Photograph by the author.

a teaspoon of topsoil and a trillion bacteria under each square foot of the surface. If you're like me—definitely not one of this country's wealthiest 1 percent—and have trouble comprehending numbers like a billion or a trillion, think of it in terms of weight. One acre of topsoil holds roughly a ton of bacteria, which equals the mass of a hefty Black Angus bull grazing on that same acre of soil. The diversity of bacterial life is especially high in woodland and grassland soils, which house several thousand different types of bacteria.

Bacteria are one-celled organisms only about a micrometer—or one-thousandth of a millimeter—in size. They are especially concentrated in the rhizosphere, a region only 1 to 2 millimeters wide surrounding the root, where they compete furiously for food with other microbes. Some bacteria called rhizobia ("rhiz" = root, "bio" = life) live in a symbiotic relationship with plants such as alfalfa, clover, beans, and peas. They infect the root hairs, forming nodules and converting nitrogen from the air into a form the host plant can use, while the plant supplies carbon compounds to the bacteria.

A group of branching filamentous bacteria called actinomycetes is responsible for the earthy smell of moist, freshly turned soil (fig. 48).

Gardeners are especially familiar with this rich odor since they work in close contact with the ground rather than in enclosed tractors. The odor is the result of a compound released as the actinomycetes break down organic matter (Lowenfels and Lewis 2010). Actinomycetes like *Streptomyces* are a source for antibiotics used in humans and animals, but for millions of years before we began using them to combat human diseases, antibiotics produced by these beneficial bacteria were faithfully controlling bacterial populations in the soil.

Under the microscope, these branching actinomycetes resemble fungi, a group of multicellular organisms that includes yeasts, rusts, smuts, molds, mildews, truffles, and mushrooms. Fungi are not plants or animals but belong to their very own kingdom. While mushrooms, as one example, are the above-surface fruit that disperses the spores of a fungus, the parts of the organism below the ground are the most remarkable. A fungus grows by sending out microscopically thin and hollow filaments called hyphae, usually less than 4 micrometers thick—which is less than one 250th of a millimeter. But the hyphae can reach enormous lengths. A handful of good forest soil might contain 10 to 20 miles of fungal hyphae! Even though the individual strands are too small to see with the naked eye, if you've ever noticed white fibrous masses in your garden soil, they were probably bundles of a few hundred thousand hyphae (fig. 49). If left undisturbed, an individual fungus can cover many acres. However, the tiny individual filaments are very fragile and easily damaged by regular tillage. When preparing samples for glass slides in our soil health lab, I learned very quickly to disperse a bit of soil in water by shaking the tube only gently so the hyphae wouldn't break up into their individual tiny cells.

Unlike plants, fungi do not contain chlorophyll and so have to depend on other ways to get their food. They specialize in converting hard-to-digest organic material such as the woody compounds lignin and cellulose, which bacteria tend to avoid, into forms that other organisms can use. By extending their hyphae out for great distances, they are able to locate new food sources and transport nutrients between locations many feet or yards apart. Bacteria generally don't move more than a few inches and require a film of water in order to do so. Fungi can extend up into the leaf litter and bring nutrients back

49. Masses of fungal hyphae on immature compost. Photograph by Zach
Wright, www.livingsoil.net.

down to the root zone; they can even penetrate hard surfaces like rock
fragments, bones, and the chitinous shells of insects.

Some of the most interesting and essential organisms in this group
and in plant life in general are the mycorrhizal fungi, which colonize
roots and live in a symbiotic relationship with at least 95 percent of
the plants on Earth. Over most of history, people considered soil fungi
harmful to plants. Then in the 1880s, the German government com-
missioned a scientist to study truffles, a highly prized foodstuff. He
discovered the important association between soil fungi and roots,
even though he never did figure out how to increase the truffle population (Pennisi 2004). Unfortunately, since most soil scientists didn't
pay a lot of attention to soil biology and ecology for many years, the sig-
nificance of these symbiotic fungi was not widely acknowledged until
the 1990s.

Mycorrhizae commonly grow on the surface layers of tree roots or
grow outward from within or between the root cells of grasses, row

crops, and most vegetables. They obtain needed carbon from the carbohydrates in root exudates while doing their critical job of bringing water and nutrients to the plant—mainly phosphorus and nitrogen but also the micronutrients copper, zinc, and iron. In essence, they extend the reach of the plant a phenomenal amount by increasing the surface area of its roots by several hundred times. So these living organisms partner with good soil structure and porosity to provide plants with all they need by exploiting water and nutrients in both the horizontal and the vertical directions. What's more, some fungi enhance soil structure by producing a sticky protein called glomalin, which coats peds as the fungal strands grow through soil pores. The glomalin acts as a soil glue, binding soil particles together and improving the stability of the peds.

Fungi can be very important agents of disease suppression. Ironically, most plant pathogens happen to be fungi, but broad-spectrum fungicides, which are toxic to mycorrhizal fungi, are not the solution. Such pesticides kill both the pest and the beneficial fungi, and agronomists have learned that only the pest species excel at building up resistance to the pesticides. On the other hand, a diverse soil ecosystem encourages beneficial organisms and keeps pest species such as smut fungi or rust fungi in check because they have to compete for resources. Abundant colonies of the delicate mycorrhizal fungi are especially important for protecting root surfaces.

Such a diverse soil ecosystem has a variety of organisms, both micro and macro, at all levels of the food web. In the microbe category, it includes the all-important protozoa and the nematodes. Protozoa are microscopic single-celled organisms, but they can be hundreds of times larger than bacteria. Amoebae are one type, along with two other groups called ciliates and flagellates. There are more than 60,000 known protozoan species in nature, of which the majority live in the soil (Nardi 2007). A teaspoon of good topsoil contains several thousand individual protozoa.

Protozoa are especially active in the rhizosphere around roots, where they feed primarily on bacteria, but they also eat other protozoa, soluble organic matter, and even fungi. Through their feeding and when they die, protozoa play a very important role in making nutrients like nitrogen available to plants. Plants need nitrogen in larger amounts than any other nutrient, but very little is present in the minerals of the soil's parent material. Instead, it originates in the atmosphere, which is 78 percent nitrogen. But plants cannot use it in its gaseous form. So the question is, How do plants get this essential macronutrient? The answer: by way of living organisms. Nitrogen must be cycled through the hierarchy of the soil food web, beginning with certain bacteria that are able to convert nitrogen in the atmosphere into plant-available forms like ammonium. As described earlier, rhizobia living on the roots of legumes are one such type of bacteria, but there are plenty of others living freely in the soil, such as *Azotobacter* (*azote* = nitrogen in French).

After the plant has taken up nitrogen in the form of ammonium or nitrate, it eventually dies and becomes part of the soil organic matter. During decomposition of the organic matter, bacteria employ enzymes to break down the bonds holding organic chains together. In the process, proteins are converted to plant-available forms of nitrogen. Since the average life span of a bacterium is only a few minutes to a few hours, it is when a protozoan eats a bacterium (dead or alive) that the real magic happens. The protozoan needs about thirty times more carbon than nitrogen, and the bacterium contains only about five times more carbon than nitrogen, so the protozoan has to eat a lot of bacteria to meet its carbon quota. In the process, it gets more nitrogen than it needs, so it releases the excess nitrogen into the soil where it can be used by plants (Magdoff and Van Es 2009).

This explanation may seem a bit academic, but it's at the very heart of the role that living organisms play in the soil. Their part in the nitrogen cycle is one example of how microbes in the soil have teamed up with plants for millions of years to the benefit of both, a process that humans have been trying to circumvent for decades now. The cost and environmental damage associated with the annual application of billions of pounds of fertilizers stand in stark contrast to the naturally enriching and regenerating capacity of soil organisms as they simply live out their lives.

There is another important way in which nutrient cycling contributes to the sustainability of the ecosystem. The food web immobilizes nitrogen when plant growth has slowed, such as in late summer, and retains it in both the living organisms and the dead organic matter. As a result, this organic nitrogen is less likely to be lost from the root zone compared to inorganic nitrogen from manufactured fertilizers, which is usually in the form of nitrate. A nitrate molecule is extremely mobile because it has a negative charge and will not attach to negatively charged clay or organic matter. As a result, inorganic nitrogen from fertilizers is easily lost to groundwater or surface runoff and is the source of Iowa's huge problem with nitrate contamination of drinking water.

After bacteria, protozoa are the dominant microbe in most soils. But nematodes are another important group. These nonsegmented worms often get a bad rap for causing plant disease, even though most nematodes are beneficial to plants. There are five general groups, based on their diet, four of which are free-living in the soil. Some feed on bacteria, some on fungi, some prey on protozoa and other nematodes, and others are omnivores. The fifth group is made up of parasites that attach to and feed on roots. It is only these root-eating nematodes that growers need to be concerned about, and they constitute only a small minority in healthy soils with a diverse biology.

Similar to protozoa, beneficial nematodes release nutrients in plant-available forms. They can also regulate the growth of prey populations such as bacteria or other nematodes and are a food source for larger soil organisms like arthropods, which include soil insects. There can be up to several hundred nematodes in a teaspoon of soil—more in forest soils, fewer in grassland soils, and even fewer in agricultural soils. They are more common in coarser-textured soils, for instance, sandy loams, because of their relatively large size. Like protozoa, they can move only through the soil water (fig. 50).

Arthropods ("arthros" = jointed, "podos" = legs) are invertebrates that range in size from microscopic to several inches in length. In the soil, they include mites, spiders, millipedes, centipedes, crayfish, and numerous insects such as springtails, beetles, termites, earwigs, ants, and many others, including the larval forms of some butterflies and moths. Several dozen different species may live in a square mile of row crops, while a square mile of forest soil could be host to several

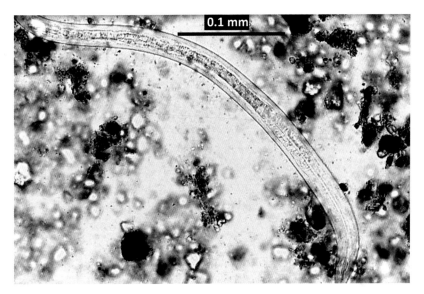

50. A bacteria-feeding nematode, with its mouth at the bottom of the photo. Photograph by the author.

thousand species. Most eat bacteria, fungi, worms, or other arthropods. This stimulates the growth of microbe populations, including the highly beneficial mycorrhizal fungi. As arthropods feed, they aerate the soil by creating channels and pores of many sizes. Some also shred organic material on the soil surface, and on agricultural fields where insufficient dead plant residues are left on the surface, they may become pests by feeding on live roots.

Another important role of arthropods and other small soil fauna is the cycling of soil particles through their guts. When soil passes through the gut of an arthropod or an earthworm, as nearly all healthy topsoil eventually does, it gets intimately mixed with organic matter. When excreted as fecal pellets, this rich substance is a concentrated nutrient resource for bacteria and fungi. In addition, larger pellets can become numerous enough over time for a fine granular soil structure to develop, with the granular peds generally measuring less than 2.5 millimeters in diameter. They are coated with organic compounds from the gut and may also be smeared with clay in fine-grained, clay-rich soils. Peds coated in this manner can be extremely stable. Recall from chapter 4 the strong granular structure in the A horizon of the Pleistocene paleosol from Adams County. It remains remarkably intact despite being overrun twice by hundreds of feet of ice and buried by some 70 feet of dense glacial till and other sediments.

In our brief catalog of soil organisms, we've finally reached the domain of earthworms. Humans have marveled over these native tillers of the soil for centuries. Earthworms do not move through existing pores like many organisms but make their own openings, greatly increasing the porosity of the soil. They often plow 7 to 8 feet into the soil, partly pushing and partly eating their way through it. Being very sensitive to ultraviolet light from the sun, earthworms stay below-ground to eat. They are eating machines who get their nutrition from the microbes—especially bacteria—growing on organic matter in the soil they consume. The rest of the organic matter gets recycled into the soil through their wastes. Night crawlers venture aboveground at night to harvest plant debris, which they pull down into their burrows and shred, thereby mixing organic matter deep into the soil profile and making it available for other organisms.

As soil passes through an earthworm's gut, it becomes enriched with the nutrients needed by plants. The waste products or casts contain around 50 percent more calcium, nitrogen, phosphorus, potassium, and bacteria than the surrounding soil. Charles Darwin observed the activities of earthworms throughout his lifetime; from studies of prehistoric and Roman ruins in the British Isles buried over the centuries by earthworm casts, he estimated that earthworms added as much as 40 tons of casts each year to each acre of soil.

The family of earthworms most important to the health of agricultural soils is Lumbricidae, which originated in Europe but which human activities transported to many parts of the world, including North America. Acknowledging their role in plowing the land long before humans existed, Darwin wrote, "It may be doubted whether there are many other animals which have played so important a part in the history of the world, as have these lowly organized creatures" (1881).

When my colleague Rick Bednarek gave soil health presentations to groups around the state, one of his most effective slides simply showed a shovel holding freshly exposed topsoil with the caption "the most important scientific instrument you'll ever own." He encouraged farmers and master gardeners to dig into their soil and notice how many earthworms were living in that shovelful of soil, because earthworms are the best and most visible indicators of a healthy soil ecosystem (Staudt 2019). In the spring of 2013, Iowa soil scientist Neil Sass conducted a simple study in Bremer County by digging five small soil pits in a no-till field and five pits in an adjacent cultivated field in the same soil type. He found an average of 29.4 earthworms living in a cubic foot of soil in the no-till field—ranging from 21 to 40 worms across the five samples—but an average of only 1.4 earthworms per cubic foot in the cultivated field—ranging from 0 to 4 worms—just 200 feet away. In other words, the no-till field had at least twenty times more worms doing what they've been doing for millennia: mixing the soil with organic matter and plowing and aerating the soil. In contrast, in Neil's words, tillage subjects a soil's inhabitants to something "akin to an earthquake, tornado, and fire happening all at the same time."

Beside the esteemed earthworms, numerous other types of invertebrate animals live in the soil, including some that are nearly microscopic like the rotifers, which move through water films in the soil with a whirling crown of cilia, and larger ones such as snails and slugs. The soil is also home to an amazing variety of vertebrates—reptiles, amphibians, mammals, and birds—who make their homes in it and circulate air and minerals through it. Some of them move huge amounts of soil, and although some can be pests by feeding on roots and plants, most prey on other soil animals. Most people have seen examples of all of these but are unaware, as was I, of the great diversity among these giants of the soil. The most readable yet authoritative guide I've seen for anyone interested in exploring the surprising variety of organisms in the soil is *Life in the Soil: A Guide for Naturalists and Gardeners*, written and beautifully illustrated by University of Illinois biologist James Nardi.

Clearly, life in the soil is diverse and fascinating, from the smallest bacteria to the largest badger. So how do we grow our food, whether on farms or in gardens, in a way that respects this abundance and diversity of life? How can we team up with soil organisms as our vital partners rather than dismissing them as interesting but superfluous creatures in the ground? The next chapter illustrates how some Iowa growers are already doing this by practicing sustainable farming through different kinds of soil health practices. Also encouraging are the many groups and organizations in our state that have jumped on the bandwagon and are promoting the adoption of sustainable farming practices.

CHAPTER EIGHT

Stories from the Field

The soil has no vote nor voice but ours. Who will speak for the dumb acres? . . . I speak up for the right of a soil to exist, and not be consigned to abuse and even to extinction.
—FRANCIS D. HOLE, 1993

I'VE HEARD IT SAID that when a soil dies, it is no longer soil but becomes the four-letter word "dirt." We cannot afford to continue growing our food in dirt doctored with synthetic chemicals, more and more tons of which are needed as the soil becomes more degraded and devoid of life each year. Recognizing life in the soil and understanding the central role that soil organisms and organic matter play in its well-being are the first steps toward reversing this trend and finding a path not only to sustainable but to *regenerative* agriculture that will restore health to Iowa's degraded soils. The next step is figuring out just how to feed the soil and protect this valuable ecosystem on millions of acres of Iowa cropland. For horticultural activities such as gardening, the same principles apply, simply on a smaller scale.

The good news is that more and more conscientious and courageous farmers and landowners in Iowa and around the Midwest are finding innovative ways to do just that. Regenerative farming and gardening are not one-size-fits-all undertakings, because different soils and plots of land call for different approaches. Managing land for soil health requires thought, creativity, and trial and error on the part of the grower. Gabe Brown from North Dakota, who farms for soil health and has written a book about it, claims that "if you don't have any failures, you're probably not trying hard enough." The intent of this chapter is not to provide a step-by-step recipe that fits everyone, an impossibility, or to examine in detail the pros and cons of various farming practices or management options. There are numerous other resources where growers can find ideas and guidance, since many groups and organizations in the state have made it part or all of their mission to promote soil health. These range from organizations doing research to those offering farm field days around the state and from peer groups to individuals providing one-on-one assistance in the field. In this chapter, I will explain the important principles of managing for soil health through regenerative agriculture by highlighting a few farms where Iowa's new pioneers are actively improving their soils.

What we are learning is that with proper care, many soils are surprisingly resilient and can be made healthy once again. For that to happen, though, important changes must take place. Soil biology, soil structure, the stability of the peds, and a soil's porosity must improve and organic matter must increase. Together, these lead to enriched fertility, better infiltration of rainfall, and much less erosion and polluted runoff.

The very foundation of healthy soils and regenerative agriculture is proper management of the soil's organic matter. On nearly all Iowa's cropland, where organic matter has been seriously depleted for many years now, that means managing the soil in a way that builds organic matter and begins to store carbon in the soil rather than releasing it into the atmosphere. Building organic matter is the key to developing a strong soil ecosystem that is self-sufficient, requiring only sunlight and rainfall, and self-regulating, with diverse organisms that keep pests under control. To become self-sufficient, a soil must use the sun's energy, rainfall, and nutrients efficiently by way of the natural food web, in which the primary producers—the photosynthesizing organisms such as plants—become food for a diverse assemblage of more and more complex organisms, the consumers. So diversity is the key, both above and below the ground.

Let's take a look at how some farmers in Iowa are achieving these goals by adhering to one or more principles of building healthy soils. The most important of these are planting cover crops that keep living roots in the soil for most of the year to feed the organisms, disturbing the soil with tillage and field traffic as little as possible, keeping crop residues on the surface year round to protect and nourish the topsoil, and planting diverse cash crops and diverse cover crops that support a healthy soil ecosystem.

Healthy Soils through Regenerative Agriculture

One of the most important ways to maintain the diversity of soil life is to keep roots living in the soil for most of the year to feed the organisms. Most corn fields and bean fields in Iowa contain living roots for less than half the year, generally only May through September. For the other seven months, the soil lies fallow and the organisms starve. Fungi are particularly vulnerable during this fallow period. Along with disturbance of the soil by tillage, the lack of living roots for much of the year helps explain why fungal populations are sparse to nonexistent in most row crop soils. The best way to maintain high populations of mycorrhizal fungi spores during the fallow period between cash crops is to plant cover crops, which have numerous other benefits as well.

People have used cover crops to improve soil since antiquity—for longer than 3,000 years in China, for example. The main purpose of cover crops in the past was to provide green manure to improve soil fertility. On our family farm in northern Michigan in the 1950s and 1960s, my father routinely planted winter wheat or alfalfa after harvesting a grain crop to add nitrogen to the soil for the following year's potato crop. (Potato harvest came too late for the cover-crop seed choices of the day to germinate in our northern climate.) Planting a green manure crop was a time-honored tradition brought from the old country, and it was simply the thing to do when you couldn't afford expensive fertilizers.

With the growing interest in soil health of the past few decades, cover crops are experiencing a welcome comeback in the Midwest and around the country. The 2017 Census of Agriculture indicated that Iowa farmers planted 973,112 cropland acres to cover crops in 2017, the most for any state that year except Texas (National Agricultural Statistics Service 2019). Iowa Learning Farms at Iowa State University estimated that about 888,000 acres of cover crops were planted in Iowa in 2018 (Juchems 2019). Both numbers were less than 4 percent of Iowa's cropland, but they do represent an increase of at least half a million acres since 2012. Most new cover-croppers begin small to test the waters, usually planting fewer than 100 acres of cover crops. The Iowa Farm and Rural Life Poll, which ISU Extension has conducted annually since 1982, showed that 23 percent of respondents used cover crops in 2017 on some of their land (Arbuckle 2019). This was double the number of farmers in 2010. Changes in the 2018 farm bill regarding cover crops and eligibility for federal crop insurance may have removed one impediment to adoption of the practice by more farmers.

Beside maintaining living roots through what were once fallow periods and adding organic matter and nitrogen to the soil, cover crops are very effective at preventing erosion. The living vegetation aboveground breaks the destructive impact of raindrops by providing a protective roof over the soil and also slows any minor runoff that might develop. The living roots belowground improve soil porosity and soil structure, increasing the infiltration of rainfall and drastically reducing or eliminating runoff. Exudates from living roots are an important glue that holds the peds together. Roots also hold on to the soil and prevent the detachment of soil particles, which often carry pollutants such as excess phosphorus with them. As a result, cover crops are an excellent practice for maintaining cleaner surface water in streams, which of course is a major objective all across Iowa.

An additional benefit of cover crops, especially for organic growers, is their capacity to break pest cycles and suppress weeds. Because organic growers cannot depend on pesticides to do that work, they have traditionally relied on tillage to combat weeds. Now, many are beginning to use cover crops as an alternative to disturbing the soil with cultivators.

Keota farmer Levi Lyle is one Iowa grower who has taken to cover crops in earnest. After earning bachelor's and master's degrees at the

51. Levi Lyle planting soybeans while terminating a cereal rye cover crop by roller crimping, which avoids soil damage from multiple field passes. Photograph by Jason Johnson, USDA–Natural Resources Conservation Service.

University of Northern Iowa, Levi worked with the Meskwaki Nation and at Simpson College before moving back to the family farm in 2012. Shortly afterward, he began planting a cover crop of cereal rye in the fall where organic soybeans would be planted in the next growing season. He allows the rye to grow tall before rolling it down with a roller crimper so that it forms a weed-suppressing mat, rather than terminating it with herbicides. Timing the crimping just right is critical: if the rye is rolled before it sheds its pollen, it bounces back up rather than slowly dying, and if it's rolled too late, its seeds drop and germinate along with the intended crop. Levi roller crimps the cover crop in May as he plants soybeans right into the living rye (fig. 51). He has also experimented with other species like Austrian winter peas and fava beans for cover crops before planting organic corn, since the decomposition of rye does tie up nitrogen and can reduce corn yields unless nitrogen fertilizer is applied. Levi's experiment has been so successful that others have taken note, and he now rents out his roller crimper, developed by a local manufacturer, to several other eastern Iowa farmers, both organic and conventional. Others have since bought their own crimpers from an Ohio manufacturer.

Levi sees cover crops as an excellent alternative to herbicides that helps reduce strain on the environment and also allows his young children to safely take part in some of the farming operations. He is inspired by neighbors who have been cover-cropping for more than twenty years and "have completely eliminated erosion on their hillsides." His parents, although more traditional farmers, are very supportive of his commitment to healthy soils and healthy food and believe that his is "the way of the future." He now farms nearly 700 acres with his father. As a member of Practical Farmers of Iowa, Levi Lyle is representative of the innovative farmers who hold the fate of Iowa agriculture in their hands. In essence, these younger farmers are figuring out the farming methods of the future, but some of their methods bear a close resemblance to those of the past before large-scale, chemical-dependent agriculture took root.

For organic farmers, a cover-crop mat is extremely beneficial to the soil because it eliminates multiple passes over the field with a cultivator to uproot or bury weeds between rows. By avoiding tillage for weed suppression, Levi is following another of the principles of regenerative agriculture—minimizing the amount of disturbance to the soil, chief among which is tillage. In addition to organic matter, healthy soil needs excellent tilth, which gives it all the physical properties necessary for good seedling emergence and root penetration. Cover crops are great for breaking up compaction and improving soil structure and porosity, but they cannot do it alone. The best way to limit disturbance to the soil is to reduce tillage before and after planting or to eliminate it altogether.

As described in chapter 6, tillage destroys soil structure and over the years creates a dense compacted layer, the plow pan. The upper several inches of the topsoil may also become compacted, creating a soil addicted to tillage, because the usual solution is more tillage to break up the soil into clods before planting. And so the cycle repeats itself, year after year.

Fortunately, practices known as conservation tillage have helped considerably in addressing compaction, at least in fields where they've been adopted. Such practices have been around for more than forty years and range from those that merely reduce tillage to the no-tillage system—commonly known as no-till. Conservation tillage became possible with the development of herbicides that could kill weeds chemically rather than mechanically and machinery that could plant seeds into soil covered by the previous year's crop residues. It caught on because it saved time, fuel, and money, but it also conserved soil structure and protected the soil from erosion by leaving much of the crop residues on the surface throughout the year. Under no-till management, the planter merely cuts a planting slit through the residues and into the soil. Cultivating for weed control between rows during the growing season under no-till is replaced by herbicides, unless cover crops are used in tandem with no-till. In that case, the thick plant residues left on the surface from the cover crop prevent much of the weed growth, so fewer if any herbicides are needed to kill summer weeds.

According to the Census of Agriculture, one-third of Iowa's harvested cropland was farmed in 2017 using no-till, and another one-fifth was farmed under some other form of conservation tillage. Those numbers were up by 15 to 20 percent from the 2012 census. In the 2018 Iowa Farm and Rural Life Poll of about 2,000 farmers, half the respondents reported using no-till on at least some of the land they farmed in 2017 (Arbuckle 2019). Unfortunately, much of that land is only under no-till for soybeans but is tilled the next year in the rotation for corn. Iowa Learning Farms reported that further growth in the use of no-till had stalled in 2018 (Juchems 2019). Fluctuation in the adoption of farm practices is very closely tied to the markets as well as to programs and rules of the federal farm bill, so it is difficult to predict future trends. In addition, some farmers admit to enjoying tillage almost as a recreational activity, and after investing hundreds of thousands of dollars in equipment, they feel the need to use it. Of course, that's less of a motivator when crop prices are down and fuel costs are up, so in combination with increased outreach and education this too may slowly change.

Most organic growers and nearly all nonorganic cover-crop farmers employ conservation tillage. Washington County farmer Steve Berger has been using no-till for more than forty years and has never plowed his fields. In the 1970s, when Steve was a teenager, his father, Dennis, was one of the first farmers in the state to adopt no-till in a county that has been Iowa's leader when it comes to soil health practices long before they were called that. Steve grew up with no-till, got a degree from Iowa State, and added his own stamp to the legacy when he began planting cover crops around 2007. He now has cover crops of cereal rye growing on all his 2,200 acres of corn and soybeans from harvest until spring planting, so his soils are never bare and hold living roots most of the year. Like his father, he is passionate about preventing erosion on his fields. When I visited him, he waved a large mounted aerial photograph taken in the late spring, which shows his cover-cropped fields alongside two neighboring fields, one under no-till and the other conventionally tilled. While deep ephemeral gullies are clearly visible in the conventional field and have even advanced well upslope into the no-till field, they stop abruptly at the Berger property line where cover crops now hold court.

In Steve's fields, sheet and rill erosion is extremely low, and gully erosion is absent because the soil has a roof over its head and a framework of roots to support it throughout the year. The canopy of a living cover crop during the off-season and a mat of dead plant residues during the growing season together make up the roof. The plant residues are an accumulation of the dead cover-crop material plus the corn or bean stover left on the field after harvest rather than being tilled into the soil. Together, the roof and the roots provide tremendous protection to the soil from both pummeling raindrops and any surface runoff that might develop during heavy rains.

Leaving crop residues on the surface year round is another important principle of regenerative agriculture. It goes a long way toward mimicking natural conditions—where the ground is seldom devoid of vegetation—while producing crops with reliably good yields even during years of less than optimal weather conditions. On average, Steve's corn yields are 25 percent higher and his bean yields are 15 percent higher than the county averages. This despite having soils with a lower average Corn Suitability Rating than the county average. In his words, "The soil doesn't lie. You improve your soil, and you will improve your crops."

Residues on the surface also feed soil organisms in the subsurface because earthworms are superb slow-motion vacuum cleaners—relentlessly pulling vegetative matter down into their holes where they shred it and mix it with the soil. They basically use it as bait, because they then eat the organisms feeding on the shredded plant matter. Steve supplements this on-farm organic matter with manure including turkey manure, which is readily available in the area. A meticulous record keeper, Steve showed me a plot of the amount of organic matter in his soils at a few dozen points, where he tests every four years. Based on the fact that his latest numbers are between 3 and 4 percent, he estimated that the organic matter of his soils has increased by about one-tenth of a percent each year since he began cover-cropping, basically doubling over that time.

Steve is a champion advocate of soil health practices known across Iowa and nationally for his expertise. He has spoken at hundreds of farm field days and meetings around Iowa and the Midwest in the past decade. Because he walks the walk and has been so successful at it, farmers tend to listen when Steve talks. He believes in inspiring by example and is always willing to share what has worked for him and how others might tailor it to their own farms.

Cereal rye, the cover crop of choice on the Berger farm, is the most commonly planted cover crop in Iowa. Nearly 90 percent of cover-croppers responding to the Iowa Learning Farms survey used cereal rye, while 22 percent used tillage radishes and 21 percent planted oats in 2018 (Juchems 2019). Agronomists and soil health experts recommend planting a mix of cover-crop plants as insurance, since some species may not thrive under the weather conditions of any particular growing season. So diversity in both crop rotations and cover crops is yet another principle of regenerative agriculture for restoring soil health.

Multispecies cover crops, planted as seed mixtures or "cocktails" of both legume and nonlegume species, offer the best of both worlds. Legumes like peas, beans, or clovers can fix anywhere between 50 and

52. Cover crop of radishes, oats, and rye seeded after corn harvest on a Century Farm in Davis County. Photograph by Jason Johnson, USDA–Natural Resources Conservation Service.

150 pounds of nitrogen per acre for the coming cash crop. Grasses such as rye, oats, wheat, or triticale are especially good at increasing organic matter in the soil. Buckwheat grows very rapidly and is therefore an excellent choice for suppressing weeds following an early harvest. Brassicas used as cover crops include mustard, rapeseed, and radishes. Tillage radishes have thick roots up to 20 inches long—and a much longer taproot—that break through compacted layers, leaving a friable soil with improved infiltration and storage of rainwater and better root penetration for the following crop (fig. 52). Broadleaf species also have the important advantage of providing canopy and shade, keeping summer surface temperatures cooler than simple crop residues by as much as 20 degrees. Flowering cover crops such as buckwheat, clovers, and brassicas also provide a great food source for pollinators and other beneficial insects, especially at times when other farm crops are not flowering (Clark 2007).

The pros and cons of various cover-crop mixtures are many and complex. While some are well understood and well documented, many have yet to be discovered. No single mixture works best on all farms or with all cash-crop rotations, but a number of Iowa farmers are experimenting with different mixtures to find what works best for them.

Crop rotations and multispecies cover crops have been the tool of choice for Thom Miller of Henry County, as just one example. A first-generation farmer, Thom is aggressively building the soils on his fields, which were in conventionally tilled corn and beans for decades. When he acquired the property, its Clinton soils—Alfisols in a loess parent material—were severely eroded, with next to no topsoil left and organic matter content of less than 1 percent. For three years, he planted a multispecies cover crop that included cereal rye, oats, sunflowers, buckwheat, brown midrib corn, soybeans, collard greens, turnips, and radishes. Each year, he grazed his cows on the cover crop, adding manure to the mix. (Incidentally, the cows' favorite was the turnips.) Allowing controlled grazing by livestock on cover crops can be very beneficial to the soil. The manure boosts organic matter, and the roots slough off carbon-rich exudates into the soil as the plants are grazed. On the livestock end of things, cover-crop species such as turnips, wheat, and rye can provide valuable forage for cows in the fall and spring and sometimes even into the winter.

After three years of diverse cover crops, Thom planted soybeans. When harvest time came around, they yielded more than 90 bushels per acre on this previously degraded soil, well above the county average. He then planted annual rye grass and alfalfa to remain in the ground for another three to five years to further build the soils. As a result of Thom's efforts, the organic matter in these Clinton soils had visibly increased when I visited the field with Jason Steele, the soil scientist working with Thom. I could see for myself that the moist surface layer was dark grayish brown (10YR 4/2) instead of the brown (10YR 5/3) Jason had found just a few years earlier. This suggests that it contained somewhere in the range of 1.5 to 2 percent organic matter or double what it had been. Just as important, the porosity of the new topsoil was greatly improved. Thom and Jason are convinced that a diverse cover crop, especially one including plants with deep taproots, can work wonders in worn-out soils. Jason is now following a similar formula on some of his own farmland in Jefferson County, also in Clinton soils.

Remarkable diversity is the hallmark of a thriving 700-acre certified organic farm in Shelby County northwest of Harlan. Rosmann Family Farms is nestled in tightly rolling country only 20 miles east of where Iowa's Loess Hills begin their steep climb. Due to years of conservation efforts, erosion of these hilly loess soils is nowhere to be seen during my July visit. These days Ron Rosmann, his wife, Maria Vakulskas Rosmann, and their two adult sons not only grow a variety of cover crops like hairy vetch, rye, millet, radishes, and turnips, they grow diverse cash crops, too—seven varieties of corn, four varieties of soybean, several small grains, and a few acres of popcorn to boot. In fact, most years they plant at least twenty-five different kinds of seed. For his hogs, Ron plants a mix of oats, wheat, barley, and field peas, which I learn are the ingredients for his high-protein succotash feed. Beside hogs, the Rosmanns raise both a fall herd and a spring herd of Red Angus cattle on 120 acres of nourishing, rotationally grazed pasture. They have not used pesticides on their crops or antibiotics on their animals since 1983, and their farmstead has been certified organic since 1994. Between green manure from legumes like red clover and alfalfa and composted animal manure, they haven't had to apply any nitrogen since 1982!

During my visit, I am impressed by Ron's knowledge of seed varieties and seed mixes in the context of organic farming. Perhaps his biology degree from Iowa State kindled an interest in experimentation, because he tells me that he has conducted more than forty trials of different crops over the years. As we walk into a field, he relates insights he has gleaned from his years of research trials. When grown together, a crop of alfalfa and orchard grass has few pest problems compared to alfalfa alone. Defoliation of soybeans by thistle caterpillars—a stage of the painted lady butterfly—is not a concern unless 30 percent or more of the foliage is affected, even though many nonorganic growers spray pesticides at the first sign of a lacy leaf like the one Ron points out to me. Tilling results in five to seven times more weeds than practicing no-till and removing weeds with a rotary hoe, which creates very little soil disturbance. Nuggets of wisdom honed by years of experience.

When the Rosmanns first went organic, their main crop was soybeans for the Japanese tofu market. Some thirty-five years later, they sell organic corn, soybeans, grain, and meat to several well-known organic companies and have their own label as well. A few years ago Maria, who previously worked as a journalist, opened a store called Farm Sweet Farm on the farmstead where she sells their own organic beef, pork, eggs, and popcorn as well as many other products made in the area. Her store is just another ingredient in a diverse agricultural system, which the Rosmanns firmly believe imparts a stability and a resiliency to farming that are becoming increasingly important under our changing climate.

I'm not surprised to learn that Ron has given several presentations around the country on cultivating resilient crops and communities and acts as a farm adviser to a U.S. senator. A few weeks after my visit, the Rosmanns host a farm field day sponsored by Practical Farmers of Iowa. The topic, "Cultivating Farm Resilience for a Changing Climate," brings together several speakers, including Art Cullen, Pulitzer Prize–winning editor of the *Storm Lake Times*, and Iowa State's Rick Cruse. At least 115 folks attend, some towing young children as we ride hay wagons and visit the farm's succotash and bean fields, a surprisingly pleasant-smelling open hoop building thriving with nursing piglets, and a hundred-foot-long compost pile that receives all the cattle, hog, and chicken manure produced on the farm except for the grazed pastures and fields (fig. 53). Throughout the day, Ron promotes Mother Nature, who provides a wide range of "services for free," instead of relying on monoculture farming and chemicals.

Unlike the highly diversified acres where the Rosmanns plant, tend, and harvest crops and raise livestock on their own land, tenant farmers work more than half of Iowa's farmland. This central feature of Iowa agriculture is partly a consequence of federal crop insurance. Crop insurance allows operators without equity in landownership to obtain the loans needed to farm large areas. It is also a huge factor affecting soil health. The dynamics of conservation get a lot more

53. The Rosmanns' compost pile in Shelby County, one stop on a Practical Farmers of Iowa farm field day. Photograph by the author.

complicated when the landowner isn't doing any of the actual farming and often isn't even paying much attention to what's happening in the fields. Many landlords live several counties away or even out of state, and too often their only concern is adjusting cash rental rates to reflect corn and soybean yields. As of 2018, 40 percent of all rented farmland in Iowa was owned by people who had never farmed and in many cases didn't even live on the land they owned (Worley 2018). Among the many interesting findings of a 2018 Iowa State survey is the fact that 41 percent of farm operators who were renting land reported that their landlords had invested zero dollars in conservation during the previous decade (Zhang, Plastina, and Sawadgo 2018). If you do the math, that adds up to more than 5 million acres of farmland and soils left to the tenant farmers to protect. Some conscientious tenants do invest their own dollars in conservation methods and cover crops on land they don't own, but most do not.

The survey also indicated that 19 percent of landlords lived out of state and 40 percent of landlords were women. One landowner who fits both these descriptions is Ruth Rabinowitz, who lives in California and oversees farmland spread over six Iowa counties. Similar to many of the state's landowners who came to manage land following the death of a spouse or parent, she recently assumed ownership of the farms from her elderly father, who loved the Iowa and South Dakota land he had purchased over the years. Never having lived on a farm, Ruth was thrust headlong into an overwhelming milieu of leases, crops, agrochemical terminology, farming practices, soil health options, and the acronym world of conservation programs. She initially considered hiring a farm management agency but ultimately decided that, as an artist, she would make shaping these acres "the art project of my life." She sought out all the help she could find from the Natural Resources Conservation Service, the Iowa Farmers Union, Practical Farmers of Iowa, the Xerces Society, the Environmental Defense Fund, the Women, Food and Agriculture Network, and other sources.

Taking over from her father, who had visited Iowa annually, Ruth found she needed to do some housecleaning. While residing in Iowa for two and a half months, she met with all her farm operators, replaced some of them, and began creating detailed farm plans with her tenants. This is a work in progress, and Ruth is very committed to maintaining a holistic relationship with the land and her tenants for the long term. Her properties already have a number of grassed waterways, wetlands, conservation buffers, Conservation Reserve Program (CRP) plantings, a pollinator planting, and some cover crops, but she is working on adding more cover crops and a prairie planting for pheasant habitat. When I met her, she was hosting an Iowa Farmers Union event on her land in Madison County, where a dozen or so landowners, operators, and representatives from various groups listened to a presentation about the Natural Resources Conservation Service's resource stewardship tool and engaged in lively discussion of the landlord-tenant working relationship. It was a supportive setting in which farm folks, both men and women, could learn from one another and brainstorm about future steps toward responsible farming.

Managing Grazing for Healthy Soils

As Ruth recognizes so well from the patchwork of her own acres, farming is often about much more than growing corn and beans. Up until now, this chapter has focused on soil health on cropland, but it is important to point out that cultivated land isn't the only place in Iowa where soil degradation is taking place. Another very significant resource in the state is its 3.3 million acres of pastureland, a million acres of which are prime farmland, and the many additional acres of woodland that are grazed mostly by beef cattle and some dairy herds. Pasture conditions in Iowa are of course a good deal better than in many parts of the country, simply due to our state's grass-friendly climate. However, soils in many of Iowa's pastures are in decline due to poor grazing practices, compaction, and the gully erosion commonly seen along cattle trails.

When I worked in southern New Mexico, I was shocked at the condition of most of the grazing land during summers with less rainfall than normal, where "normal" is only 8 to 10 inches a year. Seeing emaciated cows walk a few hundred feet in search of an isolated tuft of grass was pretty demoralizing for this midwesterner. So when I moved back to Iowa, the verdant pastures of our southern counties visible from I-35 were like eye candy. I thought they must be heaven, a bovine field of dreams. I began to think a little differently about it when I learned more about what continuous grazing does to the grass and to the soil. When cattle are given unrestricted access to the same pasture all summer long, the grass never has a chance to grow tall or reach maturity, so the roots never extend very deep into the ground. As a result, the soil organic matter decreases over time. In addition, just like us, cows tend to be choosy about what they eat and will selectively graze on the lushest forage. As a result, some plant species begin to wither, manure nutrients become concentrated in certain areas rather than distributed throughout the pasture, and soil compaction from cattle hooves degrades the soil. In hilly terrain like much of Iowa, grass kept short by grazing allows runoff to collect and erode deep gullies where once there were cow paths.

Producers can choose from a number of prescribed grazing strategies to improve both forage and soils, including rotational grazing, strip grazing, mob grazing and others (Flack 2016). These strategies all have in common the fact that they give the grass a chance to recover and deepen its roots. Cattle are rotated among several paddocks separated by movable electric fencing on a prescribed schedule. Some farmers think of this as drought insurance, because the improved forage with deeper roots can withstand dry periods much better than shallow-rooted pastures. In addition to better water infiltration and retention, which reduce erosion from runoff, prescribed grazing builds soil organic matter, improves soil respiration, and helps control weeds. It also allows for an extended grazing season and more plant diversity, which produces healthier animals that are not as dependent on grain supplements. The "grass farmers" who practice it have found those benefits to be more than enough to offset the additional labor and fencing it requires.

From 2015 to 2019, agency soil scientists and I collected topsoil samples in hopes of seeing how grazing practices affect the soil over time on both macro and micro scales. One example was a comparison of the same Sharpsburg soil type in two pastures in Union County—one continuously grazed and the other under rotational grazing. The two soils showed an obvious physical difference in the degree of compaction. The continuously grazed soil broke out in thick plates or layers as we dug into it with spades, and its higher bulk density was apparent as we carefully carved out 5x7-inch "undisturbed" samples with dirt knives for the preparation of thin sections. The view under the microscope validated what our hands were telling us. While the continuously grazed soil showed the horizontal cracks common in compacted soils, the soil under prescribed grazing revealed a predominance of vertical pores. These allow for much better infiltration of rainfall, making the soil more tolerant of drought.

Another grazed pasture, this one in Hamilton County, tells an inspiring story of Iowa initiative. Lost Lake Farm is a small dairy and cheese-making operation north of Ames run by Ranae and Kevin Dietzel. Raised on a farm near Radcliffe, Iowa, Ranae eventually obtained advanced degrees in soil science at Cornell University and sustainable agriculture and agronomy at Iowa State University. Kevin lived on a dairy farm in Minnesota for much of his childhood; after graduating from high school, he did a farming apprenticeship on three diversified dairy farms in southern Germany, becoming a state-certified agriculturalist. He followed this unconventional but valuable beginning with a biology degree from the University of Minnesota and several years in a wide variety of agricultural experiences. The Dietzels' backgrounds prepared them for their farming enterprise in a way that would have been unheard of forty or fifty years ago, when today's farmers of retirement age were getting their start. But a college education is no longer uncommon among Iowa's young farmers. Thankfully so, because farming is no longer your grandparents' operation. Ultramodern farming techniques like real-time kinematic satellite positioning systems, which can navigate field rows on the sub-inch scale, are becoming standard fare on the back forty. However, regenerative farming in a rapidly changing climate and market environment presents new challenges that also require good old-fashioned ingenuity and low-tech hard work to go along with technological advances.

To meet these challenges, the Dietzels practice prescribed grazing and soil health strategies religiously. They rotate their small but growing herd of Brown Swiss and Normande milk cows from paddock to paddock twice every day after trips to the barn for milking. Lost Lake Farm sits on the former north shore of a shallow glacial lake drained in 1895, and their part of the old lake bed is dry and grazeable most summers. They've also created pasture area out of former corn and soybean fields by seeding them with sorghum-sudangrass, followed by forage oats with an underseeding of legumes and pasture grasses. Although their operation is not certified organic, they use no synthetic chemicals on their land and use antibiotics very rarely. They are a perfect example of organic-ish farming. In other words, not USDA-certified but ecologically conscientious farming that takes good care of the soil and is very close to being organic. They strive for high-quality milk and healthy, happy animals rather than high milk production. The couple does well selling their homemade cheeses online and at farmers markets in Ames and Des Moines. Organic-ish farming like this, which is becoming more common, works especially well for small family-owned farms, which consumers feel they can trust. People tend to believe the word of dedicated farmers like the Dietzels when they say, "We follow our own rigorous standards in line with our goals to improve the land we are on . . . and provide pure and simple delicious cheese."

One of the pastures at Lost Lake Farm, which sits on a summit above the old lake bed, has been grazed for the past several years but has never been cultivated. Sampling the soil there, we found a good 24 inches of black topsoil in the A and AB horizons! Despite having a parent material of dense clay loam till once lodged beneath a massive glacier, this Mollisol of the Bode soil series showed no compaction and exhibited beautiful strong crumb structure—a type of granular structure that has less-rounded peds (fig. 54). I had a difficult time believing that this soil was in glacial till, but texture doesn't lie. Compare the porosity of this healthy soil to that of the tilled and compacted soil in figure 45, chapter 6.

Properly managed grazing can be a valuable enhancement to soil health because of the fertilization by manure. Many consider it to be a fifth principle of soil health and regenerative agriculture. The average cow can produce over 50 pounds of manure and urine every day, nearly 9 tons per grazing season. If poorly managed, of course, manure in pastures can cause water-quality problems. If it is managed well under an effective grazing system, however, the cattle will spread the manure around the pasture evenly so nutrients do not become concentrated where they drink and rest. Unlike in a crop field, nutrients in a pasture are recycled daily—from plants as livestock consume them, then back

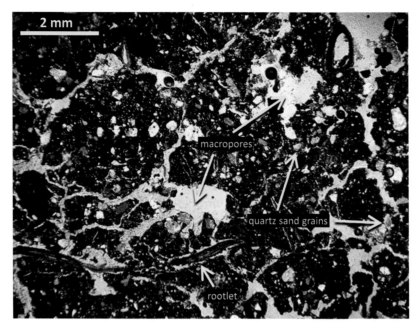

54. Superb crumb structure and porosity in the topsoil of rotationally grazed pasture at Lost Lake Farm, Hamilton County. Photograph by the author.

into the soils in the form of urine and manure. Over time, good grazing practices can result in a large increase in that all-important soil ingredient, organic matter, and the improved soil structure and overall soil health that come with it. One downside, of course, is the large volume of greenhouse gas methane that livestock produce, mainly by belching.

Soil Health on Urban Land

So far I have focused on agriculture, but Iowa has yet another soil environment very worthy of attention. Of our state's 1.9 million acres of developed land, roughly one-third is classified as urban land. Take away the buildings, paved streets, and parking lots, and most of the rest consists of private and commercial lawns, parks, and golf courses, presumed to be permeable ground surface. However, like crop fields, very few urban soils are models of good soil health, and the 64 percent of Iowans who live in cities and large towns—according to the latest available census data of 2010—would be wrong to think themselves somehow greener than the state's rural residents when it comes to their soils. This is especially true in newer housing and commercial developments, where grading of soils during construction removes all or almost all the topsoil and results in compacted soils very low in organic matter. Sometimes the contractor replaces the soil with whatever fill dirt is close by, since the industry philosophy is "dirt is dirt." Consequently, grass establishment is often poor, soil structure is lacking, and water infiltration is also poor. For instance, in 2007, when I moved into a newly rehabbed home in one of the long-established neighborhoods of Des Moines with towering oaks, I was dismayed to find that the contractor had stripped away all the topsoil down to the dense clay loam of the B horizon, so compacted that I couldn't push a soil probe into the ground more than 2 inches. In these situations, many people simply pile on the chemicals to counteract the soil's lack of fertility and the weeds that thrive in that kind of environment.

However, and more importantly, beyond homeowners' or business owners' immediate concerns about the appearance of their lawns, there are serious issues with off-site water pollution from urban run-off as a result of heavy chemical use. According to the National Research Council, suburban lawns and gardens on average receive more pesticides per acre than the average amount applied to agricultural fields (Beyond Pesticides 2017). Along with impervious pavements and huge parking lots, urban lawns, parks, and golf courses contribute a substantial share of the runoff that transports fertilizers, pesticides—especially herbicides like 2,4-D—and sediment to nearby water bodies, including heavily used recreational rivers and lakes. Although smaller in quantity than the off-farm sediments and other pollutants delivered by cropland runoff, chemicals from urban runoff in densely populated parts of the state may affect more Iowans than agricultural contaminants do. It's a growing problem for many Iowa cities and towns, because research has shown that urban growth of even 1 percent fuels negative biological impacts on surface waters.

Some of Iowa's soil and water conservation districts, including those in Polk County and Johnson County, have been addressing the issue of urban runoff and stormwater management for several years in partnership with municipal governments. As a first step, they encourage residents and businesses to conduct soil-quality restoration on their lawns as a way of reducing urban runoff and the pollutants it carries to water bodies. The objectives of soil-quality restoration are to reduce soil compaction, increase pore space, and increase organic matter, with its sponge-like ability to absorb rain. In other words, improving the health and functionality of the soils, not that different from the soil health objectives of agriculture. Practices include composting, aeration, seeding, and deep tillage where absolutely necessary on densely compacted soils (Polk Soil and Water Conservation District 2016).

In 2019, the Polk district in partnership with four state entities rolled out an exciting new initiative called Rainscaping Iowa, which offers at least a dozen practices designed to increase infiltration of water into the ground and reduce runoff to the stormwater system. Replacing turf and landscaping with deep-rooted native grasses, which greatly increase soil porosity and organic matter, can radically improve the health of soils over large expanses of lawn in parks, in yards, and on campuses (fig. 55). Other practices include rain gardens and permeable pavements. The city of Des Moines offers cost-share funding to homeowners who follow stormwater best management practices with the goal of improving the water quality of nearby streams and lakes by reducing urban runoff (Iowa Storm Water Education Partnership 2019, 2020). In June 2018, a historic flash flood in Des Moines painfully brought home the need for these practices. When several stormwater sewers experienced backups and many streets flooded, police received multiple requests for rescue. Floodwaters swept away two members of my church—one of whom drowned—and damaged or destroyed some 2,000 homes.

Often all it takes to convince people of the enormous difference between a healthy, porous soil and a dead, compacted one is a simple demonstration using two beakers and a couple of clods of soil. I saw this test performed many times when I traveled the state with Rick Bednarek to talk at farmer and gardener events. The demo almost

55. Root systems of nonnative Kentucky bluegrass (left) with turfs of native blue grama (middle) and buffalo grass (right). Roots of these native species, shown at mature height, extend down to at least 6 feet. Illustrations by Doug Adamson. Courtesy of the Polk Soil and Water Conservation District.

always met with surprise. When you pour an equal amount of water onto two intact soil clods, one with good crumb structure and the other lacking structure due to tillage and compaction, all the water seeps through the first clod in a matter of a minute or so, while most of the water is still standing on the second sample an hour later. To see this for yourself, watch the YouTube video "RaytheSoilGuy Demonstrates the Infiltration Test." In the field, much of that water at the surface becomes runoff—recall figure 46. Even though the YouTube demonstration uses cultivated soils, the principle is the same. It doesn't matter whether farm machinery and tillage or urban construction equipment caused the compaction.

In addition to the problem of runoff from urban lawns, compacted soils can present a real challenge to gardeners. But the principles that work for regenerative agriculture apply just as well to gardening and other horticultural operations, simply on a smaller scale. As with agriculture, there is no single recipe for gardening for soil health, but gardens can be ideal laboratories for discovering what works.

When Rick and I talked about soils with groups of master gardeners, he often told the story of his wife's gardening experience at the home they purchased in 2011 in a brand-new subdivision. Unable to penetrate the rock-hard soil with a garden tiller, Beth resorted to a potato fork to break up the top few inches just enough to add some sheep and horse manure she bought locally. The following summer, she began experimenting with a cover crop of tillage radishes, turnips, and winter peas on her small plots. In the years since, the cover crop has become more and more diverse as she has added things like forage sorghum, crimson clover, safflower, the brassicas canola and collard, and any other "winter-kill" species she can find locally in a given year. Because Iowa's harsh winters kill the cover crop, she doesn't need to apply an herbicide in the spring to terminate it before planting vegetables. Beth's experience is just one example of how gardeners can follow the soil health principles of keeping living roots in the soil to feed the microbes, letting the roots and the earthworms do the tillage, maintaining plant diversity, and leaving residues on the surface when the plants die off. As a result, the Bednareks' soils have vastly improved structure and ped stability, with much better infiltration and retention of rainwater. So their gardens require significantly less watering during dry summers and produce healthful, chemical-free food.

Talking with people around the state, I've often asked, How can the soil health movement gain steam, and what will it take for more Iowa farmers and homeowners to come on board? Many of them have thought about this long and hard and offered interesting insights. One farmer told me that the science of soils and agriculture has come a long way, that we basically know what needs to be done, but that social scientists and policy makers need to figure out a way to motivate people. Another believed that it will become widespread only when the big ag companies figure out a way to make money from it, in other words, when going green becomes lucrative. One said that although he would prefer soil health practices to be voluntary, he was becoming skeptical that voluntary will ever be enough. One cover-cropper would like to see farmers given carbon credits for adding biomass to the soil. A nonorganic corn grower who terminates his cover crops with herbicides was concerned about the increasing resistance of weeds to glyphosate, better known as Roundup, but hopeful that it might eventually lead more farmers to experiment with alternative methods of cover-crop termination.

Many farmers hesitate to share their opinions with their neighbors. Like most of us, they wish to fit in with their communities, where conventional farming is often deeply entrenched. They often find that the language of yields is the only language spoken, even though slightly lower yields with cover crops usually represent a bigger profit because of lower inputs. And, in fact, a 2016–17 national survey found that cover-crop users reported an average increase of 1.3 percent in corn yields and 3.8 percent in soybean yields, although it usually takes a couple of years to see the increase. The survey elicited great enthusiasm for soil health benefits among cover-crop users, especially regarding the long-term benefits, and found a strong desire for more information and training on cover crops among nonusers (Conservation Technology Information Center 2017).

While interviewing Iowa's new pioneer farmers who are on the front lines of the soil health revolution, I was struck by the sense of responsibility they feel for their land in this, the heartland of American agriculture. They truly believe that it is critical for farmers to accept and fulfill their responsibility to America's future generations by conserving what is left of our precious soils and working toward restoring what we have damaged. They are genuinely committed to producing healthful food and healthy animals on robust land in a sustainable way. But there is no single roadmap to this goal. Meeting that challenge is a balancing act each farmer must fashion in a way that reconciles the needs of the environment, the crops, the pocketbook, and the family—and all this in a climate that is changing rapidly. On a farm, every decision involves risk, and as springs have become wetter and summer precipitation less predictable, windows of opportunity for planting and harvesting are much less reliable. Rather than putting all their eggs in one basket, some farmers are finding that diversity is the best safeguard against failure.

One cause for optimism is the exciting way these folks are sharing their ideas, successes, and temporary failures with hundreds of others at meetings and farm field events all around the state. Sponsored by organizations like Practical Farmers of Iowa, they are working together to fine-tune the system and to inspire more farmers to jump on the bandwagon. I came away from my meetings with them feeling a deep sense of respect for what they are doing. Their commitment, creativity, and risk taking are extremely admirable, although insufficiently recognized. Agriculture may be more important at our present crossroads of history than ever before. In this country, farmers themselves will be—and already are—affected by climate change to a degree that most other Americans will not be. At the same time, farmers growing crops and grazing animals are also in a position to be a major part of the solution to the crisis. Gardeners, landscapers, and homeowners have a responsibility as well. All our decisions together will determine the future of our land and its soils.

CHAPTER NINE

Soils, Climate Change, and the Future

How many names on the county clerk's plat will be there in fifty years?
I might as well try to will the sunset . . . to my brother's children. The land
belongs to the future. We come and go, but the land is always here.
—WILLA CATHER, 1913, *O PIONEERS!*

IOWA'S FARMERS are experiencing the effects of climate change more and more every year. Regenerative agriculture is critically important now, not only to reverse the damage to soils already caused by decades of conventional farming methods but to provide farms and rural communities with a degree of resilience to the increasing unpredictability of the weather compared to just a few decades ago.

The Effects of Climate Change on Farming

The National Climate Assessment, released in 2014 and again in 2018, contained extensive information about the impacts of climate change on agriculture and rural communities in the United States. Its summaries of current effects and its soundest predictions for the future are disturbing, to say the least. Since about 1980, climate change has had direct impacts on crop yields and livestock production and indirect impacts due to increased pests and pathogens. One of the biggest concerns over greater weather unpredictability is the increased uncertainty of production and its economic implications for markets and prices. This is already having overwhelming consequences for small farmers and young farmers in Iowa, who are most likely to be constrained by financing and credit. Farm resilience depends on the ability to survive a year, even two, of low production or catastrophic loss. It's no coincidence that some farm organizations are again publishing suicide hotline numbers.

"Weather" refers to the short-term variations in conditions over hours to weeks, while "climate" is the weather of a place averaged over a period of time, usually thirty years or longer. Many climate studies have focused on average changes over the past several decades, such as the increase in the global average temperature, but it's the extremes of weather within a growing season that determine successes and failures for farmers. Projections are for an increase in the number of extremely hot days and nights and for periods of severe and longer-lasting drought, alternating with periods of more intense rainfall. The timing of rainfall is also critical. The common hope among midwestern farmers has always been for a dry spring and a wet summer, which are conducive to timely planting followed by robust growth. Instead, the Midwest has been experiencing wetter springs and drier summers for the last two decades.

Between 2001 and 2011, there were three fewer workable days—days in which farmers could work their fields—in April and May than during the previous twenty years (Walthall et al. 2012). In both 2008 and 2019 in Iowa, there were only twenty-one workable days out of the sixty-one days in April and May (National Agricultural Statistics Service 2019). With the window for fieldwork narrowing, farmers are being forced to work their fields when they are too wet, further degrading the soils by compaction. Delayed planting of crops results in reductions in yields. In 2008, 2013, and again in 2019, many flooded and water-logged fields simply could not be planted at all, prompting the U.S. Department of Agriculture to offer a "prevented planting" program to partially compensate farmers for their losses. In 2013, at least 700,000 acres of cropland in northwest Iowa alone went unplanted (Takle and Gutowski 2020). In 2019, nearly 500,000 acres in Iowa were too wet to be planted (Jason Johnson, personal communication). It's not only farmers who have suffered the consequences of flooded fields; another adverse effect has been the installation of more subsurface drainage tile on farms, resulting in even more nitrogen being delivered to Iowa's water bodies (Rogovska and Cruse 2011).

Future weather unpredictability is the one thing that is predictable. However, once the crops are planted, the effects of climate change on their growth and productivity are more difficult to predict because there are so many interacting and counteracting factors. These include carbon dioxide levels, temperature, solar radiation, and precipitation patterns, the last three of which control the availability of soil water for plants. Temperature alone is not a straightforward issue, and average daily temperatures definitely do not tell the whole story. In the central United States, higher humidity, overcast skies, and increased rainfall have actually led to a so-called warming hole compared to the rest of the country—an imbalance that is expected to reverse by the end of this century. However, nighttime temperatures in Iowa and across the Corn Belt have increased more rapidly than daytime temperatures, and the number of hot nights is projected to increase by as much as 30 percent (Takle 2011). Hot nights during the grain-filling period increase the rate and decrease the duration of grain filling, which reduces yields (Hatfield et al. 2014).

On the other hand, warmer spring temperatures toward the end of the twentieth century initially allowed farmers to begin planting earlier than they had previously (Takle and Gutowski 2020). What's more, higher temperatures and carbon dioxide levels have benefited plant growth. However, an increased concentration of carbon dioxide in the air favors several weed species, which are genetically more diverse, over crop species. Exacerbating this situation for nonorganic farmers, glyphosate is less effective on weeds at higher carbon dioxide levels, necessitating higher rates of herbicide application (Hatfield et al. 2014). Seasonally, warmer temperatures in the fall and early winter allow soil microorganisms to cycle nutrients later into the year when crops no longer need them, which can increase soil nutrient losses. This also allows some pests to overwinter, leading to more generations each year and increasing the need for insecticides on nonorganic farms.

More intense precipitation, especially in the spring, increases soil erosion and the delivery of sediment and other pollutants to streams and lakes. Models based on a Cedar River Basin study suggest that over the next couple of decades, the rise in extreme precipitation events in Iowa that began around 2000 will be 2.5 to 3 times the nearly constant value for the twentieth century (Anderson, Claman, and Matilla 2015; Takle and Gutowski 2020). Another study has shown that climate change could increase the erosive force of precipitation in the United States by as much as 58 percent, so the adoption of conservation practices will be more important than ever to water quality in Iowa (Nearing 2001). The rate at which conservation practices will be adopted and the degree to which they will mitigate the predicted increase in erosion due to intense rainfall are uncertain.

Ironically, those same practices are also desperately needed to moderate the effects of drought, the other side of the double-edged sword of climate change threatening farmers. In 2012, 100 percent of Iowa experienced severe drought conditions and 65 percent of the state suffered extreme drought. Since then, portions of the state have experienced severe or extreme drought at different times, most recently in 2018 and 2020. These alternating years of drought and flood—weather whiplash—that Iowa has experienced in the past twenty-five to thirty years have exacerbated nitrogen levels in streams and lakes, which is of concern to all Iowans (Loecke et al. 2017).

Under conventional tillage systems, drought leads to crusting and breakdown of soil structure and cracking to depths of 20 inches or more. Since the only way for plants to take up soil nutrients is through the soil water, there is an excess of nitrate in the soil (Jerry Hatfield, personal communication). In addition, aerobic soil microbes become increasingly inactive due to high temperatures and lack of water. Conversely, the diverse root systems provided by cover crops and crop rotations foster a much wider range of microbes compared to conventional farming methods, increasing the chances of nutrient cycling. No-till methods, cover crops, and increased crop residues also improve the infiltration of scarce rainfall, build the organic matter needed to hold precious water, and moderate soil temperatures (Al-Kaisi and Kwaw-Mensah 2017; Al-Kaisi et al. 2013).

In 2018, drought severely affected livestock producers in six counties in southeast Iowa, where grazing is extensive. The federal government declared Davis County a natural disaster area that summer due to extreme drought. Under such conditions, sources of water for pastured livestock dry up and forage is greatly diminished. Pastures with compacted soils are especially vulnerable because their grasses are shallow-rooted and unable to reach deeper soil water even if it is available. The little rain that might fall is lost to evaporation before it can infiltrate into the compacted ground.

Faced with such extremes, farmers must be adaptable in order to survive. Healthy soils will help buffer climatic extremes. Increasing the amount of organic matter and improving the soil structure of cropped and grazed soils are the two most influential changes farmers can make. Both improvements greatly increase the infiltration of water into the soil and the soil's ability to hold on to more of that water later into the growing season. Besides, halting the loss of carbon caused by tillage and instead growing the carbon pool with practices like cover crops, better crop rotations, and prescribed grazing are beneficial for another urgent reason.

Sequestering Carbon in Soils

The topic of carbon storage in soils has become an increasingly important part of the climate crisis discussion since the 1997 Kyoto Protocol provided for the creation of a carbon sequestration market. Capturing carbon was subsequently a major emphasis of the Paris Climate Agreement of 2015 (from which the U.S. is in the process of withdrawing as I write this). Sequestering carbon means keeping it stored in plants and soils in order to keep it out of the atmosphere, where it joins with oxygen to form carbon dioxide. (The term "carbon" in this discussion refers specifically to soil organic carbon.)

Carbon dioxide and methane are the most detrimental of the greenhouse gases responsible for raising the average global temperature over the past several decades. Carbon credit markets have existed in some European Union countries since 2002, and the concept is gaining traction in the United States as one part of the solution to the climate crisis. Recently, a number of articles have appeared, both nationally and in Iowa, promoting the role farmers can play in sequestering carbon. Art Cullen of the *Storm Lake Times* maintains that most Iowa farmers do not consider government bailouts a solution and would be willing to change to more sustainable practices if they were paid for environmental services like sequestering carbon to help offset the costs of implementing these practices. Matt Russell, a Grinnell farmer and director of a religious nonprofit dedicated to climate issues called Iowa Interfaith Power and Light, believes that the government could do this at a fraction of the cost taxpayers are currently shouldering for farm bailouts to offset losses from flooding, drought, and trade tariffs (Leonard and Russell 2019).

In reality, instead of investing in regenerative agriculture, the government is subsidizing agricultural failures. Federal farm policy also rewards overproduction, when it could be rewarding farmers for maximizing environmental benefits instead of yields alone. It's also possible that many farmers already are capturing more net carbon than their operations produce and thus have a smaller carbon footprint than those of us who are purely consumers. They simply aren't getting paid for their service (Barth 2019). Although politics may be largely responsible for this, another reason is the extreme difficulty of accurately monitoring and tracking changes in soil carbon.

There is no doubt that soils can store a massive amount of carbon. Our planet's soils contain more than 3 times as much carbon as the atmosphere and about 4.5 times as much as all the plant and animal life on Earth! Amazingly, those tiny but mighty mycorrhizal fungi are a major carbon storage mechanism, holding almost 30 percent of the soils' carbon (Lowenfels and Lewis 2010). Improved management to bring mistreated land back to health could certainly sequester additional carbon, although the potential for soils to store more carbon is finite. Theoretically, soils could sequester additional carbon equal to the amount they have lost over historic times. For example, it is likely that the carbon lost from the conversion of grassland to cropland can be gained back by reversing the process, that is, converting cropland to grassland (Guo and Gifford 2002). However, given the food needs of a rapidly growing global population—expected to reach nearly 10 billion people in the next forty years—only about half the theoretical amount of additional carbon is considered attainable globally within forty to fifty years, even under best-case scenarios of land use and management (Lal 2004).

In addition, the balance sheet for net carbon storage must take into account modern agriculture's ongoing contribution of carbon to the atmosphere. In the United States, this comes mainly from the combustion of fossil fuels, decomposition of soil organic carbon due to tillage, drainage of wetlands, methane from digestive fermentation in livestock, and nitrous oxide from nitrogen fertilizers. Together they contribute roughly 10 percent of total U.S. emissions of greenhouse gases, although estimates vary. Taking such agricultural emissions into account, a 2004 estimate based on the best research available at the time showed that carbon sequestration by soils worldwide has the potential to offset 5 to 15 percent of global fossil fuel emissions (Lal 2004). Similarly, the United Nations' Intergovernmental Panel on Climate Change in 2019 put the estimate at 4 to 12 percent of the current emission rate, roughly 10 gigatons per year. A less optimistic projection put the potential offset of fossil fuel emissions by improvements in soil management, including biochar application, at no more than 5 percent (Schlesinger and Amundson 2019). (Biochar consists of wood or other organic material burned under conditions of low oxygen to prevent it from going to ash; such charcoal is a very stable form of carbon.) Others have concluded that beneficial changes in cropland and livestock management could provide about 7 percent of the climate mitigation needed to meet the Paris agreement's goal of holding global warming below 2 degrees Celsius through a combination of conservation tillage, cover crops, diverse rotations, and nutrient management (Griscom et al. 2017). So the science is still imprecise.

Even though their potential contributions are difficult to quantify, improvements in agriculture can clearly make a difference to the world's carbon balance. But these alone are not a magic solution to the buildup of greenhouse gases, as some have suggested. Slowing the buildup of greenhouse gases will require a concerted effort by all sectors of society. However, because more than 80 percent of Iowa consists of agricultural land, farmers and livestock producers in Iowa and a few other Corn Belt states may be in a position to contribute to global carbon sequestration to a disproportionate degree—giving them not only an opportunity but a responsibility.

There are other excellent ways to sequester carbon in Iowa. One is through the protection of wetlands. Per unit area, wetlands hold more carbon than either forests or grasslands and much more than croplands. One study estimated that wetland restoration in the formerly glaciated prairie pothole region—the Des Moines Lobe in north-central Iowa, large parts of Minnesota, North Dakota, and South Dakota, and much of south-central Canada—has the potential in ten years to offset 2.4 percent of the annual fossil carbon dioxide emissions reported for North America in 1990 (Euliss et al. 2006). And people are beginning to recognize the value of wetlands, albeit slowly, particularly with respect to water quality and waterfowl and invertebrate habitats. Wetlands have extremely diverse plant and animal communities and are very proficient at removing contaminants from the groundwater filtering through them (Loynachan et al.). So avoiding the further loss of our remaining wetlands and restoring many that were drained in the past would have many benefits in addition to sequestering carbon.

Prairie restoration is another tremendous opportunity for capturing carbon here in Iowa, more so perhaps than in any other state. The simple fact that more than two-thirds of Iowa's soils were Mollisols is undeniable proof that prairie plants thrive here. Given our future unpredictable climate, deep-rooted prairie species are an ideal choice for the job because they love our hot summers and are able to withstand a wider range of rainfall conditions, including drought, than any other class of midwestern vegetation. For years there has been a lot of emphasis on planting trees to address global warming, but compared to forests, much more of a prairie's biomass turns into carbon-rich humus in the soil (Krug and Hollinger 2003). Best of all, after the initial planting, prairie restoration contributes very little in the way of fossil fuel emissions to the carbon balance sheet because perennial plantings don't require annual cultivation.

All prairie plants and soil organisms work together to pull carbon out of the air and draw it down into the extensive root zone, but some plant communities may be more effective than others. At a thirty-three-year-old restored prairie in southern Illinois, the communities with the greatest diversity of grasses and flowering plants have maintained the best carbon sequestration rate over decades (Ampleman, Crawford, and Fike 2014). In central Iowa, two studies of Mollisols on prairie sites reconstructed from cultivated cropland in Jasper and Warren Counties found that rates of carbon sequestration varied significantly with slope position and, to some extent, with plant community as well (Guzman and Al-Kaisi 2010a). Interestingly, the highest rate of carbon sequestration at these sites occurred in the first ten years, the same time period during which most of the formation of new soil structure took place in the former plow layer of these degraded soils (Guzman and Al-Kaisi 2010b). This suggests that the carbon protected within soil peds may be more important for carbon sequestration than the addition of new plant materials to the topsoil. As I have pointed out so many times in this book, protecting and building soil structure are critical to regenerative land management.

There has been considerable study of the demise of soil organic carbon in the plow layer following the conversion of grassland to cropland, with loss estimates averaging 50 percent (De et al. 2020). However, before tallgrass prairie is first converted, as much as 80 percent of the soil organic carbon lies in the 4 or 5 feet below the plow layer in the deeper root zone (Krug and Hollinger 2003). Much less is known about what happens to the carbon at that depth, but we do know that fertilization under conventional agriculture promotes shallow rooting, which impoverishes carbon deeper down. Prairie restoration is an excellent way to restore carbon to the subsoil, and putting cropland acres permanently into the Conservation Reserve Program with perennial cover can achieve much the same thing over time, particularly at higher topographic positions (De et al. 2020).

Resilient Farms and Rural Communities

When more carbon is sequestered on farms, it doesn't just help reduce greenhouse gases, it also contributes to the resilience of those farms and their rural communities. It is no secret that many small towns in Iowa are dying or have died, due in large part to the demise of small and midsize family farms. As family farms go out of business, rural populations decrease, schools close or merge, local economies suffer, and businesses shut down. The majority of owners of small farms also hold jobs at local businesses, so when off-farm jobs are lost, family incomes simply can no longer sustain the considerable costs of maintaining a farming operation.

As I know from childhood, farming has always been a risky business, hinging on the weather, pests, and the markets. The derecho in August 2020 that damaged or destroyed millions of acres of corn in Iowa with hurricane-level winds is one extreme example. When you add in the increased weather-related risks that climate change brings, it seems very likely that this trend will only worsen without a widespread transformation of business as usual. But it isn't all doom and gloom. For years, the sustainable agriculture community across the United States has recognized and used the practices needed to make farming operations more resilient to climate change (Lengnick 2015, 2018). Widespread use of these is needed now, not just to sustain our damaged soils but to rebuild them. The group Practical Farmers of Iowa, which has been effectively promoting environmentally and economically sustainable agriculture since the 1980s, puts on farm field day events around the state that draw several hundred people (Bittman et al. 2019). More than a dozen other organizations in the state have similar objectives, including the Leopold Center for Sustainable Agriculture, Iowa Learning Farms, the Iowa Farmers Union, the Midwest Cover Crops Council, and the Women, Food and Agriculture Network, to name just five. So there are plenty of opportunities for education and peer interaction and mentoring for those wanting to learn more.

The Iowa farmers and livestock owners highlighted in the previous chapter are only a small handful of those already using regenerative practices to rebuild their soils, including cover crops, diverse crop rotations, prescribed grazing, and reduced off-farm inputs of fertilizers and pesticides. Such an agricultural operation—diverse, flexible, and self-reliant, with healthy soils that can withstand weather extremes and produce healthful food—is a more resilient business than business as usual. It can be a fulfilling endeavor for those committed, hardworking, and commendable people who farm the land responsibly and who ultimately benefit the rest of us and our environment in very important ways.

Most experts agree that the science is in place, but we urgently need social science and policy to follow suit in order for regenerative agriculture to spread and become an economically viable priority. Furthermore, we must stop considering the year-by-year health of our state's agricultural economy as justification for long-term damage to its soils and water (Connerly 2020). This is critically important to all Iowans, both rural and urban, but it is also imperative in a wider sense. Food security in a world expected to reach nearly 10 billion people in forty years will hinge on many factors, of course, but agriculture founded on healthy soils will undoubtedly be one of the most important (DeLong, Cruse, and Weiner 2017; Olson et al. 2018).

The farmers and landowners I talked with in the course of researching this book all believed in a mission beyond their own gain. We can only hope that more people in this state with its bountiful soils will come to realize that whether we own the title to a piece of land, we each have a moral responsibility to be good stewards of its resources. Legal documents come and go, but the land endures. We are only temporarily borrowing it from all the generations of people and other living creatures to come, long after we have returned to the soil. More than any of the other problems we face, the climate crisis demonstrates the fact that our individual actions have consequences for all life on Earth, in the present as well as in the future.

A second-century story tells of a group of people traveling in a boat. One of them begins drilling a hole into the boat beneath himself. When the others ask what he is doing, the man replies, "What concern is it of yours? Am I not drilling under my own place?" And they say to him, "But you will flood the boat for us all!" In its annual journey around the sun, Earth is the vessel carrying us all: people, animals, and plants. With the effects of climate change and the swell of global population barreling down on us, healthy soils will only become more and more essential, globally and here at home. Let's stop compromising our soil resources when we need them the most and begin honoring the fertile treasure bequeathed to this land we call home.

Glossary

agriculture, regenerative Farming systems that rebuild or revitalize degraded soils, biological habitats, and other natural resources through conservation and *soil health* practices

A horizon The surface layer of a soil, usually showing an accumulation of *organic matter*, biological activity, and loss of materials such as clay; often called the *topsoil*

Alfisol The order in *Soil Taxonomy* comprised of soils formed under forest or woodland vegetation characterized by a thin *A horizon*, a *Bt horizon*, and often an *E horizon*

alluvium Sediment deposited by flowing water in streambeds and on floodplains, fans, and deltas

Anthrosols Soils formed or profoundly modified by long-term human activities

Ap horizon Upper, plowed portion of the *A horizon* in many cultivated fields; may make up the entire *A horizon* in some places

B horizon Soil *horizon* beneath the *A* or *E horizon* characterized by blocky or prismatic *soil structure* and an accumulation of clay and other minerals leached from above; in the *soil profile*, it constitutes the *subsoil*

Bk horizon *B horizon* containing concentrations of *calcium carbonate*

Bt horizon *B horizon* with a significant accumulation of clay, usually in the form of coatings on the surfaces of *peds* and channels

bulk density Weight of dry soil divided by its in-place volume, usually expressed as grams per cubic centimeter

calcareous soil Soil containing enough *calcium carbonate* to visibly effervesce when treated with dilute hydrochloric acid; most Iowa soils were calcareous when they were young

calcium carbonate The mineral calcite, which is finely disseminated throughout Iowa's *parent materials* of *loess* and *glacial deposits*; *leaching* removes it from the upper soil *horizons*

carbonates, pedogenic White masses, nodules, concretions, and coatings that precipitate from the dissolved *calcium carbonate* previously brought to the *subsoil* or the *C horizon* by *leaching*; their abundance generally increases with the age of the soil

carbon sequestration Removal of carbon dioxide from the atmosphere through absorption by plants, which then add organic carbon to the soil when they die

casts Nutrient-rich droppings of earthworms; abundant casts can develop into granular *soil structure*

cation exchange capacity The potential of a soil to hold nutrients for plant use; specifically, the amount of negative charge available for the exchange of positively charged atoms and molecules (cations), both those in solution and those held by *organic matter*

chisel plow Tillage implement with narrow chisel points that rip and stir the soil but do not invert it or pulverize it as much as a *moldboard plow*; leaves more crop *residues* on the surface

C horizon Bottom portion of the *soil profile*, minimally altered by *soil formation* and usually having the same *soil texture* as the *parent material* from which the *A horizon* and *B horizon* formed

conservation buffers Small strips or areas of land in permanent vegetation, designed to intercept pollutants by slowing *runoff* and trapping sediment; they include *grassed waterways*, riparian buffers along streambanks, and *prairie strips*

contour farming System of laying out row crops so that farming operations are performed approximately parallel to the contours of a slope rather than up and down the slope; prevents rainwater from flowing downhill between rows

cover crops Grasses, legumes, or other flowering species planted after cash-crop removal for *soil health*, weed suppression, and other benefits; multi-species cover crops are the most effective

crop rotation System of growing different crops on the same piece of land in a planned recurring sequence of years; in addition to row crops, a good rotation might include alfalfa, small grains, hay, or winter *cover crops*

Des Moines Lobe Landform region in north-central Iowa created by an ice lobe of the *Laurentide ice sheet*; it surged into Iowa at least three times during the last *Wisconsinan glacial stage*

drainage class Prevailing wetness condition of a soil related to how quickly water is removed without artificial drainage; described as *poorly drained* or *well-drained soils* or several other gradations in between

drainage tile Perforated pipe or tubing laid beneath the ground surface in *poorly drained soils* to hasten the removal of subsurface water and transport it to a waterway; made of clay before the 1970s, now plastic

drought Period of unusually dry weather that persists long enough to cause extensive damage to crops and grazing land as well as water supply shortages

E horizon Occurs between the *A horizon* and the *Bt horizon* in many forest soils; usually light-colored and often silty due to the *translocation* of its clay, iron, and other oxide minerals to the *B horizon*

Entisol The order in *Soil Taxonomy* comprised of very young, weakly developed soils forming on very recent geomorphic surfaces under any type of vegetation; consists of only an *A horizon* over a *C horizon*

ephemeral gullies Shallow ditches, usually wider than they are deep, that form in fields each year and are obscured by tillage operations; they form where *runoff* concentrates in natural flow paths

expansive soil Soil that significantly changes volume upon wetting and drying due to a high content of expandable clays such as smectite clays

exudates Soluble sugars, amino acids, and other compounds secreted by plant roots

farm bill Primary U.S. legislative tool governing agricultural and food programs; renewed every five years or so

feldspar Second most common mineral in the silt fraction of most Iowa soils; it consists of the chemical elements silicon, oxygen, and aluminum plus potassium or sodium-calcium

fungi Multicelled organisms that do not photosynthesize and are neither plants nor animals; they form long chain-like filaments called *hyphae* and may form fruiting bodies such as molds or mushrooms

glacial deposit Any geologic material derived directly from glacial ice, including several types of till as well as outwash and other glacial meltwater deposits

glacial till Unsorted, generally unstratified sediments deposited beneath an advancing glacier, at the melting front of an active glacier, or by melting out on the surface of a stagnant glacier

gleyed soil Soil condition that results from prolonged conditions of saturation and *reduction*, which produce gray, greenish gray, or bluish gray colors throughout the *soil matrix* or in *mottles*

grassed waterway Broad, shallow channel planted with perennial vegetation designed to safely move surface water downslope and prevent the formation of *ephemeral gullies*

great group Third level of soil classification in the hierarchy of *Soil Taxonomy*, beneath *order* and *suborder*

Green Revolution The introduction of high-yielding, disease-resistant seed varieties and increased use of inorganic fertilizers and irrigation to Asia and Latin America in the 1950s and 1960s; it resulted in dramatic gains in wheat and rice harvests

gully erosion Displacement of soil caused by the concentrated flow of water in natural flow paths unprotected by dense vegetation; occurs by way of both *ephemeral gullies* and deep permanent gullies

highly erodible land Areas having soils with potential *sheet and rill erosion* rates at least eight times greater than their tolerable levels or *T value*

Histosol The order in *Soil Taxonomy* comprised of soils formed in materials in which *organic matter* makes up more than about 75 percent of the volume

Holocene Epoch Second epoch of the *Quaternary Period* of geologic time, occurring between about 11,700 years ago (after the last major glaciation) and the present

horizon In the *soil profile*, a horizontal layer or zone that differs from the soil above and below it in *soil texture*, amount of *organic matter*, color, *soil structure*, consistence, or other properties; the major horizons are designated by the capital letters O, A, E, B, and C

humus Stable fraction of soil *organic matter* remaining after plant and animal residues have decomposed; dark in color and consisting of only fine particles

hydric soil Soil that formed under conditions of saturation, flooding, or ponding long enough during the growing season that it lacks oxygen in its upper part

hyphae Long chain-like filaments of cells formed by *fungi*; hyphae usually live between soil *peds* rather than in the tiny pores within *peds*

Illinoian glacial stage Interval between about 300,000 and 120,000 years ago when thick continental glaciers covered much of the Upper Midwest, including part of southeast Iowa; named for deposits first described in the state of Illinois

Inceptisol The order in *Soil Taxonomy* comprised of young soils under any type of vegetation; consists of *A*, *B*, and *C horizons* but does not have a *Bt horizon*

infiltration Movement of water from the ground surface into the soil

interglacial stage Period between glacial stages when the climate warms, vegetation grows, and *soil formation* takes place; examples in Iowa include the *Sangamon interglacial stage* and the *Holocene Epoch*, our present time

Laurentide ice sheet Massive sheet of ice that covered millions of square miles of Canada and the northern United States multiple times during the *Pleistocene Epoch*

leaching Generally refers to the dissolution of finely disseminated *calcium carbonate* in the *soil matrix* and its downward *translocation* by percolating water

loess Loosely compacted deposits of silty sediments transported and deposited by wind

Mesozoic Era Interval of geologic time between about 250 million and 66 million years ago

moldboard plow Deep-tillage implement; its basic components are the blade (share) that cuts the soil into slices and the moldboard that inverts the furrow slice to bury the sod or crop *residues*

Mollisol The order in *Soil Taxonomy* comprised of soils formed under grassland vegetation such as prairie and characterized by a thick dark *A horizon* (in noneroded soils) overlying a *B horizon*

moraine Ridge composed of till deposited at the downstream end of a glacier while it stalled in one location for a sufficiently long time

mottles Small spots or blotches different in color from the surrounding *soil matrix*; they include *redox features* and patches of color inherited from the *parent material*

mycorrhizal fungi *Fungi* existing in a mutually beneficial association with plant roots; their *hyphae* radiate into the surrounding soil to gather nutrients

nematodes Tiny—usually microscopic—unsegmented worms living in the soil

nitrate Source of plant-available nitrogen; anhydrous ammonia, the most prevalent form of nitrogen fertilizer in Iowa, undergoes several conversions in the soil before becoming nitrate

nitrogen, inorganic Occurs in non-carbon-containing compounds such as ammonia and other forms derived from anhydrous ammonia fertilizer; soluble in water and thus easily lost by *leaching* into groundwater and *runoff* to surface waters in the form of *nitrate*

nitrogen, organic Occurs in carbon-containing compounds in soil *organic matter*, where it is held tightly and protected from *leaching*; decomposition of *organic matter* slowly releases *nitrate* and ammonium, which plants can use

no-till Type of conservation *tillage* whereby a crop is planted directly into a seedbed that has not been plowed or disked since the previous crop was harvested

organic matter Part of a soil that consists of plant and animal tissues and residues at various stages of decomposition; includes fresh, active (decomposing), and stable (*humus*) fractions; composed of about 60 percent organic carbon

oxidation Chemical soil-forming process by which metallic minerals such as iron and manganese combine with oxygen, losing one or more electrons, to form yellow, orange, reddish, or black oxide minerals

paleoenvironment Conditions of climate, hydrology, *topography*, and plant and animal life at a time in the geologic past

paleosol Soil that formed on a landscape of the past; as discussed in this book, they consist of buried paleosols and exhumed paleosols, which were once buried but later exposed by erosion

Paleozoic Era Interval of geologic time between about 540 million and 250 million years ago

palimpsest soil Soil in which some of the physical properties that formed during an earlier period of *soil formation* have been overwritten but not completely obliterated by later soil-forming processes

parent material Geologic sediments (*glacial till, loess, alluvium*, etc.), weathered rock, and occasionally *organic matter* in which a soil forms over time

pedology Branch of soil science concerned with *soil formation* and soil classification

peds Units of *soil structure* formed by natural processes, such as granules, crumbs, blocks, prisms, and plates; in contrast, clods form due to tillage and compaction

Peoria Formation Deposit of *loess* that fell across Iowa between about 29,000 and 15,000 years ago; it is present at the surface in most of the state, varying in thickness from several inches to more than 200 feet

permeability Capacity of a soil to allow fluids like water or gas to move through it; compare to *porosity*

pH, soil Degree of acidity or alkalinity of a soil on a scale of 0 to 14, with 7 being neutral; usually dependent on *parent material*, annual rainfall, and vegetation type

Pleistocene Epoch First and longest epoch of the *Quaternary Period* of geologic time, between about 2.7 million and 11,700 years ago, when several intervals of continental-scale glaciation occurred in the Northern Hemisphere

plow pan Thin zone at the bottom of the plow layer, having high *bulk density* and very low *porosity* caused by compaction from farm machinery and tillage

poorly drained soils Soils that are wet at shallow depths periodically during the growing season or remain wet for long periods and usually require artificial drainage for cultivation; also see *drainage class*

porosity Portion of the volume of a soil that consists of voids—pores—filled with air or water; compare to *permeability*

prairie pothole Closed depression in glaciated terrain; the site of a *wetland* or a former *wetland* later drained for agriculture

prairie strip Small section of cropland taken out of production and planted with native prairie species in a strip along the contour; reduces *soil loss* and creates wildlife habitat

Pre-Illinoian glacial stages At least seven intervals, between about 2.7 million and 500,000 years ago, when thick continental glaciers covered most or all of Iowa

prescribed grazing System of alternately grazing and resting grasslands in an orderly sequence for the purposes of improving forage and *soil health*, decreasing soil erosion, and conserving water

prime farmland Land that has the best combination of physical and chemical characteristics for producing food, feed, forage, and fiber and is also available for such use

protozoa One-celled soil organisms much larger than bacteria that have a nucleus similar to plant and animal cells; they include amoeba, ciliate, and flagellate types

quartz Most common mineral by far in the silt and sand fractions of most Iowa soils; it consists of the chemical elements silicon and oxygen

Quaternary Period Interval of geologic time from 2.7 million years ago to the present; it includes the *Pleistocene Epoch*, the *Holocene Epoch*, and soon the proposed Anthropocene Epoch

radiocarbon dating Also called carbon-14 dating, this is a method for estimating the age of charcoal or other *organic matter* in soils or sediments by measuring a sample's remaining amount of the unstable radioactive isotope carbon-14

redox depletion Spot or zone of decreased pigmentation that is grayer and usually lighter than the surrounding *soil matrix*; a field indicator of regular saturation

redox features Concentrations, nodules, concretions, or coatings of iron or manganese oxides resulting from *oxidation*, as well as *redox depletions* resulting from *reduction*

reduction Chemical soil-forming process by which iron and manganese minerals and certain other compounds gain one or more electrons under saturated conditions, resulting in *redox depletions* or an entire *soil matrix* with grayish colors

residues, crop Plant materials such as stalks, stems, leaves, or seed pods left in the field after harvest; they contribute *organic matter* to the soil and help reduce erosion, moderate soil temperatures, and prevent evaporation

runoff Portion of precipitation that does not infiltrate the soil but rather flows off the land surface to channels; it can cause *sheet and rill erosion* and the formation of *ephemeral gullies* on exposed soils

Sangamon interglacial stage Interval between about 120,000 and 80,000 years ago after the retreat of the *Laurentide ice sheet*, when the climate in the U.S. midcontinent warmed and *soil formation* occurred across the landscape

sheet and rill erosion Detachment and transportation of soil particles by water from rainfall, snowmelt, and *runoff*

slickensides Polished, often striated faces on shear planes where one soil mass was forced past another by the pressure created during swelling of *expansive soil*

soil Ecosystem composed of mineral matter, *organic matter*, and living organisms; serves as a medium for plant growth and other functions

soil association Group of two or three *soil series* occurring together in a characteristic pattern on the landscape, typically defined by slope position or landform

soil crust Thin surface layer on soils, as much as an inch thick, that is much more compact, hard, and brittle when dry than the material beneath it

soil degradation Any process that harms soil function, including erosion by water or wind, damage to *soil structure*, loss of *organic matter*, compaction, crusting, prolonged saturation, and other processes

soil food web Community of organisms, including *soil microbes*, that live all or part of their lives in the soil

soil formation Processes by which a soil develops over time in a given *parent material* in response to the prevailing climate, organisms, and *topography*; also called pedogenesis

soil health Capacity of a soil to perform its functions; that is, to sustain biological activity, diversity, and productivity, regulate water flow, cycle nutrients, and filter or detoxify pollutants

soil loss As used in USDA procedures for estimating soil erosion, refers to the displacement of soil from its original location, which could be several feet to several miles

soil map Shows the distribution of *soil series* and their phases in relation to the *topography*, waterways, towns, and roads of an area

soil matrix Finer material of a soil, generally silt and clay in most Iowa soils, that forms a continuous substance and surrounds the coarser particles

soil microbes Generally refers to any organism too small to see with the naked eye such as bacteria, *fungi*, and *protozoa*

soil order Highest level of soil classification in the hierarchy of *Soil Taxonomy*, based on *horizon* type and development; Iowa has six of the twelve orders recognized globally

soil profile Vertical section of a soil showing all its *horizons* and extending into the unaltered *parent material*

soil series Sixth and final level of soil classification in the hierarchy of *Soil Taxonomy*; a group of soils that are similar in all major *soil profile* characteristics

soil structure Arrangement of soil particles into *peds*; described by shape, size, and strength (distinctiveness)

soil suborder Second level of soil classification in the hierarchy of *Soil Taxonomy*, beneath *soil order*

soil survey Systematic description, classification, and mapping of soils in an area

Soil Taxonomy Soil classification system used in the U.S. and many other countries, published in 1975 and significantly revised in 1999

soil texture Relative proportions by weight of sand, silt, and clay in a soil

strip cropping System of growing crops in equal-width strips, in which strips of erosion-susceptible row crops alternate with strips of erosion-resistant forages and small grains

subsoil That part of the soil lying below the *topsoil*; generally refers to the *B horizon*

terrace Several types of earthen embankments constructed across a slope to intercept *runoff* and sediment from soil erosion by water; not the same thing as a naturally formed stream terrace

thin section Very thin vertical slice of soil cut from an intact soil sample, then glued onto a glass slide for viewing under a petrographic microscope

tillage, conservation Modern operations performed for the growing of crops that reduce the loss of soil or water and leave most of the soil surface covered by crop *residues*; includes *no-till*, ridge-till, and strip-till systems

tillage, conventional Traditional operations, such as plowing or disking, performed to prepare a seedbed for crops

tilth Physical condition of a soil related to ease of tillage, fitness as a seedbed, and depth of root penetration

topography Landscape characteristics that affect *soil formation* and stability, such as type of landform, position on a slope, slope aspect—the compass direction it faces—and slope steepness

topsoil Surface *horizon* of a soil, containing more *organic matter* and microbial life than the *subsoil*; usually synonymous with the *A horizon*

transformation Chemical or physical modification of soil constituents to form other materials; for example, the weathering of minerals to form various types of clay minerals or the formation of *humus* from organic residues

translocation Movement of inorganic and organic materials within the *soil profile*, either within a *horizon* or between *horizons*, usually caused by water movement

T value Tolerable *soil loss*, defined by the Natural Resources Conservation Service as the maximum amount of a soil that can be lost annually to erosion without degrading its long-term productivity

Vertisol The order in *Soil Taxonomy* comprised of soils rich in expandable clay minerals; these soils swell and churn during the wet season and develop deep cracks during the dry season

water table Upper surface of the groundwater, below which a soil is saturated; depth to the water table fluctuates with the seasons and in Iowa is typically lowest in autumn

well-drained soils Soil in which water leaves readily but not rapidly and is available to plants in humid regions during much of the growing season; also see *drainage class*

wetland Ecosystem saturated for long enough periods to produce a *hydric soil* and support aquatic plants; in Iowa, wetlands include *prairie potholes*, bogs, fens, and floodplain swamps

Wisconsinan glacial stages Intervals between about 80,000 and 11,700 years ago when the *Laurentide ice sheet* covered much of the Upper Midwest, including north-central Iowa; named for deposits first described in the state of Wisconsin

Bibliography

Alex, Lynn M. 2000. *Iowa's Archaeological Past*. Iowa City: University of Iowa Press.

Al-Kaisi, Mahdi, Roger W. Elmore, Jose G. Guzman et al. 2013. "Drought Impact on Crop Production and Soil Environment: 2012 Experiences from Iowa." *Journal of Soil and Water Conservation* 68 (1): 19A–24A.

Al-Kaisi, Mahdi, and David Kwaw-Mensah. 2008. "Impact of Tillage and Crop Rotation Systems on Soil Carbon Sequestration." Ames: Iowa State University Extension and Outreach.

—— and ——. 2016. "Building Soil Health." Ames: Iowa State University Extension and Outreach.

—— and ——. 2017. "How Drought Affects Soil Health." Ames: Iowa State University Extension and Outreach.

Al-Kaisi, Mahdi, and Mark Licht. 2005. "Tillage Management and Soil Organic Matter." Ames: Iowa State University Extension and Outreach.

Ampleman, Matt D., Kerri M. Crawford, and David A. Fike. 2014. "Differential Soil Organic Carbon Storage at Forb- and Grass-Dominated Plant Communities, 33 Years after Tallgrass Prairie Restoration." *Plant Soil* 374: 899–913.

Anderson, Christopher J., David Claman, and Ricardo Matilla. 2015. "Iowa's Bridge and Highway Climate Change and Extreme Weather Vulnerability Assessment Pilot: Final Report." Ames: Iowa State University Institute for Transportation.

Anderson, Wayne. 1998. *Iowa's Geological Past: Three Billion Years of Earth History*. Iowa City: University of Iowa Press.

Arbuckle, J. Gordon. 2019. "Iowa Farm and Rural Life Poll: 2018 Summary Report." Ames: Iowa State University Extension and Outreach, Iowa Agriculture and Home Economics Experiment Station, Iowa Department of Agriculture and Land Stewardship, and Iowa Agricultural Statistics Service.

Arbuckle, J. Gordon, and Paul Lasley. 2013. "Iowa Farm and Rural Life Poll: 2013 Summary Report." Extension Community and Economic Development Publications 26. Ames: Iowa State University Extension and Outreach.

Argabright, M. Scott, Roger G. Cronshey, J. Douglas Helms et al. 1995. "A Historical Study of Soil Conservation." Northern Mississippi Valley Working Paper 10. Washington, D.C.: USDA–Natural Resources Conservation Service.

Balco, Greg, and Charles W. Rovey II. 2010. "Absolute Chronology for Major Pleistocene Advances of the Laurentide Ice Sheet." *Geology* 38 (9): 795–798.

Baldwin, Mark, Charles E. Kellogg, and James Thorp. 1938. "Soil Classification." In *Soils and Men: Yearbook of Agriculture*, 979–1001. Washington, D.C.: U.S. Department of Agriculture.

Balfour, Eve. 1977. "Towards a Sustainable Agriculture: The Living Soil." Presentation at the International Federation of Organic Agriculture Movements conference, Sissach, Switzerland.

Baker, Richard G., Louis J. Maher, Craig A. Chumbley et al. 1992. "Patterns of Holocene Environmental Change in the Midwestern United States." *Quaternary Research* 37 (3): 379–389.

Barth, Brian. 2019. "Thou Shalt Not Till: One Man Is Trying to Fight Climate Change by Mobilizing an Unlikely Team—Iowa's Farmers." *Mother Jones* 44 (4): 16–17.

Bettis, E. Arthur, III, Daniel R. Muhs, Helen M. Roberts et al. 2003. "Last Glacial Loess in the Conterminous USA." *Quaternary Science Reviews* 22: 1907–1946.

Bettis, E. Arthur, III, Deborah J. Quade, and Timothy J. Kemmis. 1996. *Hogs, Bogs, and Logs: Quaternary Deposits and Environmental Geology of the Des Moines Lobe.* Guidebook Series 18. Iowa City: Iowa Geological Survey.

Bettis, E. Arthur, III, and Dean M. Thompson. 1981. "Holocene Landscape Evolution in Western Iowa: Concepts, Methods, and Implications for Archaeology." In *Selected Papers from the Mankato Conference*, ed. S. F. Anfinson. Occasional Publications in Minnesota Anthropology 9. St. Paul: Minnesota Archaeological Society.

Beyond Pesticides. 2017. "Lawn and Garden Pesticides: Facts and Figures." Washington, D.C. https://www.beyondpesticides.org/assets/media /documents/bp-fact-lawnpesticides.081817.pdf.

Bittman, Mark, Sam Bennett, Sarah Carlson et al. 2019. "Can Sustainable Farming Save Iowa's Precious Soil and Water?" *PBS NewsHour*, October 12.

Boellstorff, John D. 1978. "North American Pleistocene Stages Reconsidered in Light of Probable Pliocene-Pleistocene Continental Glaciation." *Science* 202 (4365): 305–307.

Bogard, Paul. 2017. *The Ground Beneath Us: From the Oldest Cities to the Last Wilderness, What Dirt Tells Us about Who We Are.* Boston: Little, Brown.

Booth, Robert K., Stephen T. Jackson, Steven L. Forman et al. 2005. "A Severe Centennial-Scale Drought in Mid-Continental North America 4200 Years Ago and Apparent Global Linkages." *Holocene* 15 (3): 321–328.

Brady, Nyle C., and Ray R. Weil. 2008. *The Nature and Properties of Soils.* 14th ed., rev. Upper Saddle River, N.J.: Pearson Prentice Hall.

Brown, Gabe. 2018. *Dirt to Soil: One Family's Journey into Regenerative Agriculture.* White River Junction, Vt.: Chelsea Green Publishing.

Brown, P. E. 1936. *Soils of Iowa.* Special Report 3. Ames: Iowa Agricultural Experiment Station Soils Subsection and U.S. Department of Agriculture.

Burras, C. Lee. 2016. "Soil Change and Productivity in Iowa's Intensely Cropped Mollisols." https://www.nrcs.usda.gov/wps/PA_NRCS Consumption/download?cid=nrcseprd1273408&ext=pdf.

Burras, C. Lee, and Rachel K. Owen. 2012. "CSR2: Soil Productivity Rating for Cropland in Iowa, USA." Ames: Iowa State University Department of Agronomy.

Clark, Andy, ed. 2007. *Managing Cover Crops Profitably.* 3rd ed. SARE Handbook Series Book 9. College Park, Md.: Sustainable Agriculture Research and Education.

Cohen, K. M., and P. Gibbard. 2019. "Global Chronostratigraphical Correlation Table for the Last 2.7 Million Years, Version 2019 QI-500." *Quaternary International* 500: 20–31.

Connerly, Charles E. 2020. *Green, Fair, and Prosperous: Paths to a Sustainable Iowa.* Iowa City: University of Iowa Press.

Conservation Technology Information Center. 2017. "Report of the 2016–17 Cover Crop Survey." West Lafayette, Ind.: Conservation Technology Information Center with the North Central Region Sustainable Agriculture Research and Education Program and the American Seed Trade Association.

Cox, Craig, Andrew Hug, and Nils Bruzelius. 2011. "Losing Ground." Ames: Environmental Working Group Midwest.

Cox, Craig, Brett Lorenzen, and Soren Rundquist. 2013. "Washout." Ames: Environmental Working Group Midwest.

Cox, Craig, and Soren Rundquist. 2018. "Polluted Runoff: A Broken Promise Threatens Drinking Water in the Heartland." Ames: Environmental Working Group Midwest.

Cruse, Richard M. 2016. "Economic Impacts of Soil Erosion in Iowa." Leopold Center Completed Grant Reports 511. Ames: Iowa State University.

Cruse, Richard M., Scott Lee, and Tim Sklenar. 2016. "Cost of Soil Erosion." Presentation at the True Cost of American Food conference, San Francisco,

Calif. https://www.slideshare.net/sustainablefoodtrust/rick-cruse-cornsoy
-systems.

Cullen, Art. 2019. "Pay Farmers to Capture Carbon." *Storm Lake Times*,
July 19.

Darwin, Charles R. 1881. *The Formation of Vegetable Mould through the Action
of Worms, with Observations on Their Habits*. London: John Murray.

De, Mriganka, Jason A. Riopel, Larry J. Cihacek et al. 2020. "Soil Health Re-
covery after Grassland Reestablishment on Cropland: The Effects of Time
and Topographic Position." *Soil Science Society of America Journal* 84:
568–586.

DeLong, Catherine, Richard M. Cruse, and John Weiner. 2017. "The Soil Degradation Paradox: Compromising Our Resources When We Need Them the
Most." *Sustainability* 7 (1): 866–879.

Doak, Richard. 2015. "Story of Iowa: From Wetland to Farmland." *Des Moines
Register*, February 28.

Dokuchaev, Vasily. 1883. *Russkii Chernozem*. Moscow: State Publishing House
of Agricultural Literature.

Effland, William R., Hari Eswaran, and Douglas Helms et al. 2005. "A Chronological History of Science for Soil Survey in the United States of America,
1899–2006." Poster. Madison, Wis.: American Society of Agronomy. https://
www.nrcs.usda.gov/wps/portal/nrcs/detail/soils/survey/publication/?cid
=stelprdb1256535.

Eilers, Lawrence J., and Dean M. Roosa. 1994. *The Vascular Plants of Iowa:
An Annotated Checklist and Natural History*. Iowa City: University of Iowa
Press.

Eller, Donnelle. 2014. "Erosion Estimated to Cost Iowa $1 Billion in Yield."
Des Moines Register, May 3.

Euliss, Ned H., Jr., Robert A. Gleason, Alan Olness et al. 2006. "North American Prairie Wetlands Are Important Nonforested Land-Based Carbon
Storage Sites." *Science of the Total Environment* 361: 179–188.

The Farm Crisis. 2013. Iowa PBS. http://www.iowapbs.org/video/story/2388
/the-farm-crisis.

Fenton, Thomas E., and Gerald A. Miller. 1982. "Soils." In *Iowa's Natural Heritage*, ed. Tom Cooper and Nyla Sherburne Hunt, 82–93. Des Moines: Iowa
Natural Heritage Foundation and Iowa Academy of Science.

Flack, Sarah. 2016. *The Art and Science of Grazing: How Grass Farmers Can
Create Sustainable Systems for Healthy Animals and Farm Ecosystems*.
White River Junction, Vt.: Chelsea Green Publishing.

Garcia, Deborah Koons. 2012. *Symphony of the Soil: A Film*. Mill Valley, Calif.: Lily Films. https://symphonyofthesoil.com/contact.

———. 2014. "Seeing Soil." In *The Soil Underfoot: Infinite Possibilities for a Finite Resource*, ed. G. J. Churchman and E. R. Landa, 75–82. Boca Raton, Fla.: CRC Press.

Gardner, David R. 1957. "The National Cooperative Soil Survey of the United States." Ph.D. dissertation, Harvard University Graduate School in Public Administration.

Gerasimova, Maria. 2005. "Classification: Russian, Background and Principles." In *Encyclopedia of Soils in the Environment*, ed. Daniel Hillel, 1: 223–226. Oxford: Elsevier.

Gershuny, Grace, and Joe Smillie. 1999. *The Soul of Soil: A Soil-Building Guide for Master Gardeners and Farmers*. 4th ed. White River Junction, Vt.: Chelsea Green Publishing.

Griscom, Bronson W., Justin Adams, Peter W. Ellis et al. 2017. "Natural Climate Solutions." *Proceedings of the National Academy of Sciences* 114 (44): 11645–11650.

Guo, L. B., and Roger M. Gifford. 2002. "Soil Carbon Stocks and Land Use Change: A Meta Analysis." *Global Change Biology* 8 (4): 345–360.

Guzman, Jose G., and Mahdi Al-Kaisi. 2010a. "Landscape Position and Age of Reconstructed Prairies Effect on Soil Organic Carbon Sequestration Rate and Aggregate Associated Carbon." *Journal of Soil and Water Conservation* 65 (1): 9–21.

——— and ———. 2010b. "Soil Carbon Dynamics and Carbon Budget of Newly Reconstructed Tall-Grass Prairies in South Central Iowa." *Journal of Environmental Quality* 39: 136–146.

Handy, Richard L. 1964. "The Seventh Approximation: A New Pedological Scheme of Soil Classification." *Soil Survey Horizons* 5 (2): 15–23.

Hanna, Mark, and Mahdi Al-Kaisi. 2009. "Understanding and Managing Soil Compaction." Ames: Iowa State University Extension and Outreach.

Hannan, Joe. 2017. "Soil pH in the Home Garden." Ames: Iowa State University Extension and Outreach.

Hatfield, Jerry, Eugene S. Takle, Richard Grotjahn et al. 2014. "Agriculture." In *Climate Change Impacts in the United States: The Third National Climate Assessment*, ed. J. M. Melillo, T. Richmond, and G. W. Yohe, 150–174. Washington, D.C.: U.S. Global Change Research Program.

Hearst, James. 2001. *The Complete Poetry of James Hearst*. Ed. Scott Cawelti. Iowa City: University of Iowa Press.

Hole, Francis D. 1993. "The Earth Beneath Our Feet: Explorations in Community." Presentation at Eloquence and Eminence: Emeritus Faculty Lectures. Madison: University of Wisconsin.

Holliday, Vance T., ed. 1992. *Soils in Archaeology: Landscape Evolution and Human Occupation*. Washington, D.C.: Smithsonian University Press.

Hoyer, Will. 2011. "Agricultural Drainage and Wetlands: Can They Co-Exist?" Iowa City: Iowa Policy Project.

Hudson, Berman D. 1994. "Soil Organic Matter and Available Water Capacity." *Journal of Soil and Water Conservation* 49 (2): 189–194.

Ingham, Elaine R., Andrew R. Moldenke, and Clive A. Edwards. 2000. *Soil Biology Primer*. Ankeny, Iowa: Soil and Water Conservation Society and USDA–Natural Resources Conservation Service.

Intergovernmental Panel on Climate Change. 2019. *Climate Change and Land: Summary for Policy Makers*. New York: United Nations.

Iowa Association of Naturalists. 1999. "Iowa Soils." Iowa Physical Environment Series, IAN-703.

———. 2001a. "Iowa Prairies." Iowa's Biological Communities Series, IAN-203.

———. 2001b. "Iowa Wetlands." Iowa's Biological Communities Series, IAN-204.

Iowa Board of Immigration. 1870. *Iowa, the Home for Immigrants: A Treatise on the Resources of Iowa*. Printed in English, German, Norwegian, Swedish, and Dutch. Des Moines: Mills and Company.

Iowa Community Indicators Program. 2020. "Rural and Urban Population." Ames: Iowa State University. https://www.icip.iastate.edu/tables /population/rural-urban.

Iowa Department of Cultural Affairs. 2020a. "Agriculture in a Global World." https://iowaculture.gov/history/education/educator-resources/primary -source-sets/agriculture-global-world.

———. 2020b. "Great Depression and the Dust Bowl." https://iowaculture.gov /history/education/educator-resources/primary-source-sets/dust-bowl.

Iowa Department of Natural Resources. 2018. "2018 Impaired Waters Map." https://programs.iowadnr.gov/adbnet/Assessments/Summary/2018 /Impaired/Map.

Iowa Pathways: Agriculture. 2019. Iowa PBS. http://www.iowapbs.org/iowa pathways/mypath/agriculture.

"Iowa State Soil: Tama Soil Series." 2020. Ames: Iowa State University Extension and Outreach. https://www.extension.iastate.edu/soils/iowa-state -soil-tama-soil-series.

Iowa Storm Water Education Partnership. 2019. "Des Moines Metro Area Stormwater BMP Cost-Share Program." https://iowastormwater.org/wp -content/uploads/2019/08/Cost-Share-Program-amvp-2.1.pdf.

———. 2020. "Rainscaping Iowa: Landscapes for Clean Water." https://iowa stormwater.org/rainscaping.

Jenny, Hans. 1984. "The Making and Unmaking of a Fertile Soil." In *Meeting the Expectations of the Land: Essays in Sustainable Agriculture and Stewardship*, ed. Wes Jackson, Wendell Berry, and Bruce Colman, 42–55. San Francisco: North Point Press.

Jones, Christopher S., Jacob K. Nielsen, Keith E. Schilling et al. 2018. "Iowa Stream Nitrate and the Gulf of Mexico." *PLoS One* 13 (4):e0195930. https:// www.ncbi.nlm.nih.gov/pmc/articles/PMC5897004.

Juchems, Elizabeth. 2019. "Cover Crop Acres Increase, but Rate of Growth Declines in 2018." Ames: Iowa Learning Farms. https://www.iowalearning farms.org/cover-crop-acres-increase-rate-growth-declines-2018.

Kemmis, Timothy J., E. Arthur Bettis III, and George R. Hallberg. 1992. "Quaternary Geology of Conklin Quarry." Guidebook Series 13. Iowa City: Iowa Geological Survey.

Kerr, Phillip J., Stephanie Tassier-Surine, Susan Kilgore et al. 2019. "Evidence for Multiple Advances of the Southwestern Laurentide Ice Sheet during MIS 3." *Abstracts of the 20th Congress of the International Union for Quaternary Science (INQUA)*, Dublin, Ireland. https://app.oxfordabstracts.com /events/574/program-app/submission/92107.

Krasilnikov, Pavel, and Richard Arnold. 2009a. "The United States Soil Taxonomy." In *A Handbook of Soil Terminology, Correlation and Classification*, ed. Pavel Krasilnikov et al., 75–95. London: Earthscan and the International Institute for Environment and Development.

——— and ———. 2009b. "World Reference Base for Soil Resources: A Tool for International Soil Correlation." In *A Handbook of Soil Terminology, Correlation and Classification*, ed. Pavel Krasilnikov et al., 47–74. London: Earthscan and the International Institute for Environment and Development.

Krug, Edward C., and Steven E. Hollinger. 2003. "Identification of Factors That Aid Carbon Sequestration in Illinois Agricultural Systems." Contract Report 2003–02 for the Illinois Council on Food and Agricultural Research.

Lal, Rattan. 2004. "Soil Carbon Sequestration Impacts on Global Climate Change and Food Security." In *Soil, the Final Frontier*, ed. A. Sugden et al. *Science* 304 (5677): 1623–1627.

Lengnick, Laura. 2015. *Resilient Agriculture: Cultivating Food Systems for a Changing Climate*. Gabriola, B.C.: New Society Publishers.

———. 2018. "Cultivating Climate Resilience on Farms and Ranches." College Park, Md.: Sustainable Agriculture Research and Education.

Leonard, Robert, and Matt Russell. 2019. "Our Small Towns Are Toppling Like Dominoes: Why We Should Cut Some Farmers a Check." *New York Times*, June 24.

Loecke, Terrance D., Amy J. Burgin, Diego A. Riveros-Iregui et al. 2017. "Weather Whiplash in Agricultural Regions Drives Deterioration of Water Quality." *Biogeochemistry* 133: 7–15.

Logan, Terry J. 1982. "Improved Criteria for Developing Soil Loss Tolerance Levels for Cropland." In *Determinants of Soil Loss Tolerance*, Special Publication 45, ed. B. L. Schmidt. Madison, Wis.: American Society of Agronomy and Soil Science Society of America.

Lowdermilk, W. C. 1953. "Conquest of the Land through 7,000 Years." Washington, D.C.: USDA–Soil Conservation Service.

Lowenfels, Jeff, and Wayne Lewis. 2010. *Teaming with Microbes: The Organic Gardener's Guide to the Soil Food Web*. Rev. ed. Portland, Ore.: Timber Press.

Loynachan, Thomas E., Kirk W. Brown, Terrence H. Cooper et al. 2005. *Soils, Society and the Environment*. Alexandria, Va.: American Geological Institute.

Magdoff, Fred, and Harold Van Es. 2009. *Building Soils for Better Crops: Sustainable Soil Management*. 3rd ed. SARE Handbook Series Book 10. College Park, Md.: Sustainable Agriculture Research and Education.

Mallarino, Antonio P., John E. Sawyer, and Stephen K. Barnhart. 2013. "A General Guide for Crop Nutrients and Limestone Recommendations in Iowa." Rev. ed. Ames: Iowa State University Extension and Outreach.

Markewich, Helaine W., Douglas A. Wysocki, Milan Pavich et al. 2011. "Age, Genesis, and Paleoclimatic Interpretation of the Sangamon/Loveland Complex in the Lower Mississippi Valley, U.S.A." *Geological Society of America Bulletin* 123 (1–2): 21–39.

McCormack, D. E., and A. H. Paschall. 1982. "The 1934 National Reconnaissance Erosion Survey." *Soil Survey Horizons* 23 (4): 13–15.

McNeill, J. R., and Verena Winiwarter. 2004. "Breaking the Sod: Humankind, History, and Soil." In *Soil, the Final Frontier*, ed. A. Sugden et al. *Science* 304 (5677): 1627–1629.

Montgomery, David R. 2008. *Dirt: The Erosion of Civilizations*. Berkeley: University of California Press.

———. 2017. *Growing a Revolution: Bringing Our Soil Back to Life*. New York: W. W. Norton.

Muhs, Daniel R., and E. Arthur Bettis III. 2003. "Quaternary Loess-Paleosol Sequences as Examples of Climate-Driven Sedimentary Extremes." In *Extreme Depositional Environments*, ed. Marjorie A. Chan and Allen W. Archer, 53–74. Geological Society of America Special Paper 370. Boulder, Colo.

Müller, Mark. 2012. "The World Beneath Your Feet: A Closer Look at Soil and Roots." Des Moines: Iowa Department of Transportation, Iowa Living Roadway Trust Fund.

Mutel, Cornelia F. 2008. *The Emerald Horizon: The History of Nature in Iowa*. Iowa City: University of Iowa Press.

Nardi, James B. 2007. *Life in the Soil: A Guide for Naturalists and Gardeners*. Chicago: University of Chicago Press.

National Agricultural Statistics Service. 2019. *2017 Census of Agriculture: United States Summary and State Data*. Washington, D.C.: U.S. Department of Agriculture. https://www.nass.usda.gov/AgCensus.

National Soil Survey Center. 2012. *Field Book for Describing and Sampling Soils*. Version 3.0. Washington, D.C.: USDA–Natural Resources Conservation Service.

Natural Resources Conservation Service. 2007. *From the Ground Down: An Introduction to Iowa Soil Surveys*. Des Moines: U.S. Department of Agriculture.

———. 2008. "Highway Guide of Iowa Soil Associations." Des Moines: USDA–Natural Resources Conservation Service in cooperation with the Iowa Department of Agriculture and Land Stewardship, the Iowa Department of Transportation, and Iowa State University.

———. 2015. *2012 National Resources Inventory: Summary Report*. Washington, D.C.: U.S. Department of Agriculture. https://www.nrcs.usda.gov/Internet /FSE_DOCUMENTS/nrcseprd396218.pdf.

———. 2020a. *2017 National Resources Inventory: Summary Report*. Washington, D.C.: USDA–Natural Resources Conservation Service and the Center for Survey Statistics and Methodology and Ames: Iowa State University. https://www.nrcs.usda.gov/wps/portal/nrcs/main/national/technical /nra/nri/results/.

———. 2020b. *Iowa Soil Health Resources*. Des Moines: U.S. Department of Agriculture. https://www.nrcs.usda.gov/wps/portal/nrcs/ia/soils/health /soil+health.

———. 2020c. *Soil Health Producer Profiles*. Des Moines: U.S. Department of

Agriculture. https://www.nrcs.usda.gov/wps/portal/nrcs/detail/null/?cid
=stelprdb1176927.

———. 2020d. *Web Soil Survey.* Washington, D.C.: U.S. Department of Agriculture. https://websoilsurvey.sc.egov.usda.gov.

Nearing, Mark A. 2001. "Potential Changes in Rainfall Erosivity in the U.S.
with Climate Change during the 21st Century." *Journal of Soil and Water
Conservation* 56 (3): 229–232.

Olson, Carolyn, Prasanna Gowda, Jean L. Steiner et al. 2018. "Agriculture
and Rural Communities." In *Impacts, Risks, and Adaptations in the United
States: Fourth National Climate Assessment*, ed. D. R. Reidmiller et al., 2:
391–437. Washington, D.C.: U.S. Global Change Research Program.

Oschwald, William R., Frank F. Riecken, Raymond I. Dideriksen et al. 1965.
Principal Soils of Iowa: Their Formation and Properties. Special Report 42.
Ames: Iowa State University Cooperative Extension Service.

Pennisi, Elizabeth. 2004. "The Secret Life of Fungi." In *Soil, the Final Frontier*,
ed. A. Sugden et al. *Science* 304 (5677): 1620–1622.

Polk Soil and Water Conservation District. 2016. "Soil Quality Restoration:
Improving Soil Health." Ankeny, Iowa. https://www.polk-swcd.org/assets
/documents/SQR2016.pdf.

Pope, John P., and Thomas R. Marshall. 2010. "Pennsylvanian Geology of
Decatur City and Thayer Quarries." In *The Pennsylvanian Geology of South-
Central Iowa*, ed. T. R. Marshall and C. L. Fields, 3–26. Guidebook GSI-086.
Iowa City: Geological Society of Iowa.

Prior, Jean C. 1991. *Landforms of Iowa.* Iowa City: University of Iowa Press.

Prior, Jean C., Richard G. Baker, George R. Hallberg et al. 1982. "Glaciation." In
Iowa's Natural Heritage, ed. Tom Cooper and Nyla Sherburne Hunt, 47–64.
Des Moines: Iowa Natural Heritage Foundation and Iowa Academy of
Science.

Retallack, Gregory R. 2019. *Soils of the Past: An Introduction to Paleopedology.*
3rd ed. Hoboken, N.J.: Wiley-Blackwell.

Richter, Daniel deB. 2007. "Humanity's Transformation of Earth's Soil:
Pedology's New Frontier." *Soil Science* 172 (12): 957–967.

Rogovska, Natalia P., and Richard M. Cruse. 2011. "Climate Change Consequences for Agriculture in Iowa." In *Climate Change Impacts on Iowa 2010:
Report to the Governor and the Iowa General Assembly*, 14–18. Des Moines:
Iowa Climate Change Impacts Committee and Ames: Leopold Center Publications and Papers 74. https://lib.dr.iastate.edu/leopold_pubspapers/74.

Ross, Earle D. 1951. *Iowa Agriculture: An Historical Survey.* Iowa City: State Historical Society of Iowa.

Rovey, Charles W., II, and Trevor McLouth. 2015. "A Near Synthesis of Pre-Illinoian Till Stratigraphy in the Central United States: Iowa, Nebraska and Missouri." *Quaternary Science Reviews* 126: 96–111.

Roy, Martin, Peter U. Clark, Rene W. Barendregt et al. 2004. "Glacial Stratigraphy and Paleomagnetism of Late Cenozoic Deposits of the North-Central United States." *Geological Society of America Bulletin* 116 (1): 30–41.

Ruhe, Robert V. 1969. *Quaternary Landscapes in Iowa.* Ames: Iowa State University Press.

———. 1975. *Geomorphology: Geomorphic Processes and Surficial Geology.* New York: Houghton Mifflin.

Ruhe, Robert V., and John G. Cady. 1967. "The Relation of Pleistocene Geology and Soils between Bentley and Adair in Southwestern Iowa." In *Landscape Evolution and Soil Formation in Southwestern Iowa.* Technical Bulletin 1349. Washington, D.C.: USDA–Soil Conservation Service.

Rundquist, Soren, and Craig Cox. 2014. "Washout Revisited." Ames: Environmental Working Group Midwest.

Sandor, Jon, C. Lee Burras, and Michael L. Thompson. 2005. "Human Impacts on Soil Formation." In *Encyclopedia of Soils in the Environment*, ed. Daniel Hillel, 1: 520–531. Oxford: Elsevier.

Schaetzl, Randall J., and Michael L. Thompson. 2015. *Soils: Genesis and Geomorphology.* 2nd ed. New York: Cambridge University Press.

Schlesinger, William H., and Ronald Amundson. 2019. "Managing for Soil Carbon Sequestration: Let's Get Realistic." *Global Change Biology* 25 (2): 386–389.

Schulte, Lisa A., Jarad Niemi, Matthew J. Helmers et al. 2017. "Prairie Strips Improve Biodiversity and the Delivery of Multiple Ecosystem Services from Corn-Soybean Crops." *Proceedings of the National Academy of Sciences* 114 (42): 11247–11252.

Schwickerath, Wayne. 2015. "Iowa Assessment Valuation Procedure for the Agricultural Class." Nevada, Iowa: Story County Assessor's Office.

Schwieder, Dorothy. 1996. *Iowa, the Middle Land.* Ames: Iowa State University Press.

Schwieder, Dorothy, Thomas J. Morain, and Lynn Nielsen. 2002. *Iowa Past to Present: The People and the Prairie.* 3rd ed. Ames: Iowa State University Press.

Simonson, Roy W. 1959. "Outline of a Generalized Theory of Soil Genesis." *Soil Science Society of America Journal* 23 (2): 152–156.

Simonson, Roy W., Frank F. Riecken, and Guy D. Smith. 1952. *Understanding Iowa Soils: An Introduction to the Formation, Distribution, and Classification of Iowa Soils*. Dubuque: W. C. Brown.

Soil Conservation Service. 1935. *Reconnaissance Erosion Survey 1934*. Miscellaneous Publication 2. Washington, D.C.: U.S. Department of Agriculture.

Soil Survey Staff. 1975. *Soil Taxonomy: A Basic System of Soil Classification for Making and Interpreting Soil Surveys*. Agriculture Handbook 436. Washington, D.C.: USDA–Soil Conservation Service.

———. 1999. *Soil Taxonomy: A Basic System of Soil Classification for Making and Interpreting Soil Surveys*. 2nd ed. Agriculture Handbook 436. Washington, D.C.: USDA–Natural Resources Conservation Service.

Staudt, Ann. 2019. "Worms, Harbingers of Healthy Soil." Ames: Iowa Learning Farms. https://www.farmprogress.com/print/386569.

Stone, Larry A. 1999. *Listen to the Land: Selections from 25 Years of Naturalist Writing in the "Des Moines Register."* Parkersburg, Iowa: Mid-Prairie Books.

———. 2000. *IOWA: Portrait of the Land*. Des Moines: Iowa Department of Natural Resources.

Stuart, Kevin. 1984. "My Friend, the Soil: A Conversation with Hans Jenny." *Journal of Soil and Water Conservation* 39 (3): 158–161.

Swaim, Ginalie. 1986. "Dry, Dusty 1936." *Goldfinch* 7: 4.

Takle, Eugene S. 2011. "Climate Changes in Iowa." In *Climate Change Impacts on Iowa 2010: Report to the Governor and the Iowa General Assembly*, 8–13. Des Moines: Iowa Climate Change Impacts Committee and Ames: Leopold Center Publications and Papers 74. https://lib.dr.iastate.edu/leopold_pubs papers/74.

Takle, Eugene S., and William J. Gutowski Jr. 2020. "Iowa's Agriculture Is Losing Its Goldilocks Climate." *Physics Today* 73: 2.

Tandarich, John P., Robert G. Darmody, and Leon R. Follmer. 1994. "The Pedo-Weathering Profile: A Paradigm for Whole-Regolith Pedology from the Glaciated Mid-Continental United States of America." In *Whole Regolith Pedology*, ed. David L. Cremeens, Randall B. Brown, and J. Herbert Huddleston, 97–117. Special Publication 34. Madison, Wis.: Soil Science Society of America.

Targulian, Victor O., and Sergey V. Goryachkin. 2004. "Soil Memory: Types of Record, Carriers, Hierarchy and Diversity." *Revista Mexicana de Ciencias Geológicas* 21 (1): 1–8.

Tassier-Surine, Stephanie, Phillip J. Kerr, E. Arthur Bettis III et al. 2018. "Redefining the Middle Wisconsin Sheldon Creek Boundary in North Central Iowa." Presentation at the Geological Society of America North-Central Section 52nd annual meeting, Ames, Iowa. https://gsa.confex.com/gsa/2018 NC/webprogram/Paper312945.html.

Tegtmeier, Erin M., and Michael D. Duffy. 2004. "External Costs of Agricultural Production in the United States." *International Journal of Agricultural Sustainability* 2 (1): 1–20.

Thompson, Mark. 2014a. "Iowa's Black Gold." Gladys Black Environmental Education Project, Red Rock Lake Association and Marion County Conservation. https://www.gladysblackeagle.org/topics/2014-topics/iowas-black-gold.

———. 2014b. "Red Rock's Dirty Secret." Gladys Black Environmental Education Project, Red Rock Lake Association and Marion County Conservation. https://sites.google.com/a/gladysblackeagle.org/gladysblackeagle/project-ideas/red-rock-s-dirty-secret.

Thompson, Susan. 2017. "Lake Panorama RIZ 20-Year Renewal Approved." *Lake Panorama Times* 49 (4): April.

Veenstra, Jessica. 2010. "Fifty Years of Agricultural Soil Change in Iowa." Master's thesis, Iowa State University Department of Agronomy.

Walker, Rudger H., and P. E. Brown. 1936. *Soil Erosion in Iowa*. Special Report 2. Ames: Iowa Agricultural Experiment Station and USDA–Soil Conservation Service.

Wall, Joseph F. 1978. *Iowa: A Bicentennial History*. New York: W. W. Norton.

Walthall, Charles L., Jerry Hatfield, and Peter Backlund et al. 2012. *Climate Change and Agriculture in the United States: Effects and Adaptation.* Technical Bulletin 1935. Washington, D.C.: U.S. Department of Agriculture.

"Washing Away the Fields of Iowa." 2011. *New York Times*, May 4.

Wessels Living History Farm. 2003–04. "Farming in the 1930s"; "Farming in the 1940s." York, Neb. https://livinghistoryfarm.org/farming-history.

Witzke, Brian J., Robert M. McKay, Bill J. Bunker et al. 1990. "Stratigraphy and Paleoenvironments of Mississippian Strata in Keokuk and Washington Counties, Southeast Iowa." Guidebook Series 10, Tri-State Geological Field Conference. Iowa City: Iowa Geological Survey and Geological Society of Iowa.

Woida, Kathleen, and Richard Lensch. 2015. "Pleistocene Paleosols and the Clarinda Soil Series in Adams County, SW Iowa." Presentation at the Geological Society of America North-Central Section 49th annual meeting,

Madison, Wis. https://gsa.confex.com/gsa/2015NC/webprogram/Paper 255224.html.

Woida, Kathleen, and Michael L. Thompson. 1993. "Polygenesis of a Pleistocene Paleosol in Southern Iowa." *Geological Society of America Bulletin* 105 (11): 1445–1461.

Worley, Sally. 2018. "Who Owns the Farmland?" Ames: Practical Farmers of Iowa. https://practicalfarmers.org/2018/11/who-owns-the-farmland.

Zhang, Wendong, Alejandro Plastina, and Wendiam Sawadgo. 2018. *Iowa Farmland Ownership and Tenure Survey, 1982–2017: A Thirty-Five Year Perspective.* Ames: Iowa State University Department of Economics, Center for Agricultural and Rural Development, and Extension and Outreach.

Index

The letter "f" following a page number denotes a figure.

A horizon, 29f, 29–31; of Alfisols, 45, 74, 82; Ap horizon, 31, 70, 79f, 82; erosion of, 44, 70–71, 95, 106; formation of, 67, 74, 96; of Mollisols, 3, 12f, 44–45, 78, 79f, 88; in paleosols, 95, 98, 101f; thickness and crop yield, 133. *See also* topsoil
actinomycetes, 151f
Adair soil series, 97
aeration of soil: and soil drainage, 24, 148; for soil health, 182; and soil organisms, 135, 157, 159
Afton, Iowa: deep cores, 91–92, 93f
agricultural chemicals. *See* fertilizers; herbicides; inorganic nitrogen; pesticides
agricultural history of Iowa: in 1800s, 107–110; 1900–1950s, 111–114; 1950s–present, 114–118
Alfisols, 42, 43f, 44–46, 73–75, 82; profile, 45f, 46f, 83f
alluvium: soil parent material, 47, 51, 99–100
ammonium: in nitrogen cycle, 146, 155
anhydrous ammonia. *See* inorganic nitrogen
Anthrosols: defined, 105–106
Aquolls, 80, 88; defined, 42–43

archaeological use of paleosols, 86–87, 98–99, 99f
Argiaquoll, 88
argillic horizon: defined, 44
Argiudoll, 43–44
Armstrong soil series, 89–90; profile, 97f, 97–98
Arnold, Richard, 40
arthropods, 156–158
artificial drainage: on Des Moines Lobe, 81; drainage ditches, 110, 111, 111f; drainage tile, 110; and water pollution, 189

B horizon, 29f, 30–31, 73, 76; BE horizon in Alfisols, 45, 45f, 46f; Bg horizon, 78–79, 79f; Bk horizon, 32, 33f; Bss horizon, 34; Bt horizon, 31, 44, 45f, 46, 74; Bw horizon, 78, 79f; in paleosols, 95, 97f, 98
bacteria, 21, 150f, 150–152, 155, 156
Balfour, Eve, 141
Bayer (company), 117. *See also* glyphosate
Bednarek, Rick, 144, 159, 182, 184
bedrock, 29f, 62, 64, 73, 75
Bennett, Hugh Hammond, 120
Berger, Steve, 168–169
Bettis, Art, 100

biochar: defined, 193
blocky structure, 20–21, 22f, 73, 75, 76
Bode soil series, 179
Brown, P. E.: *Soil Erosion in Iowa*, 120; *Soils of Iowa*, 39
Brunizems, 39
Bucknell soil series, 89–90, 94
bulk density: defined, 27; dynamic soil property, 63
Burras, Lee, 52

C horizon, 29, 29f, 34–35, 71f, 80; of Entisols, 47, 47f, 70; in urban soils, 106
carbon dioxide: greenhouse gas, 134, 191, 194; plant growth, 150, 189
carbon emissions from agriculture, 193
carbon as plant nutrient, 145
carbon sequestration by soils, 162; carbon credits, 191; estimates of potential, 192–193; prairie restoration, 194–195; wetland protection, 194
cation exchange capacity, 146
Cedar River Basin study, 190
channelization, historic, 110
chelates, 146
chemical fertilizers, 6, 115, 140, 141. *See also* inorganic nitrogen
Chernozems, 39
chisel plow, 114, 117, 134
Clarinda soil series, 88–90; profile, 89f
classic gullies, 128–129, 176–177
clay coatings, 30–31, 32f, 75f, 76; in paleosols, 96, 97f, 98

clay in soils: in Bt horizon, 31, 32f; influence on soil behavior, 17; soil texture, 14–18, 17f, 27. *See also* expandable clay
climate: defined, 188; effect on soil color, 22; as a soil-forming factor, 68–69
climate change effects on farming: delayed and prevented planting, 188; drought, 190–191; extreme rainfall events, 189–190; interacting factors, 189; pests and pathogens, 189; weather unpredictability, 188
Colo soil series, 13
color. *See* soil color
compaction: depth of, 136; from grazing, 176–178; from tillage, 22, 135–138, 136f, 166, 183; susceptibility rating, 54
composting, 153f, 174f, 182
concretions: defined, 77
conservation compliance, 125, 130–131
conservation practices, 120–121, 130, 163–176; and global warming, 193; on rented cropland, 175. *See also* composting; contour farming; cover crops; crop residues; crop rotation; grassed waterways; prairie strips; prescribed grazing; regenerative agriculture; riparian buffers; strip cropping; terraces; windbreaks
Conservation Reserve Program, 125–126, 195
conservation tillage, 1 67. *See also* no-till

contour farming, 121

conventional tillage, 6, 135, 137f, 190; contribution to global warming, 134; description, 134. *See also* chisel plow; compaction; moldboard plow

Coppock soil series, 46–47; profile, 46f

corn, 5; and cover crops, 168, 185; history of in Iowa, 109–117; hybridization of, 113, 114; organic, 165, 172, 173; prices, 113, 115, 126; yields, 11, 113, 115, 116, 119, 133, 168, 185

Corn Suitability Rating, 52; of soils formed in paleosols, 90. *See also* CSR2

cosmogenic dating, 92; radiometric ages, 93f

counties, Iowa. *See* Iowa counties

county extension offices: creation of, 112

cover crops, 163–166, 168–172, 170f, 175, 190; to build organic matter, 162, 169, 170; for forage, 178–179; terminating, 165, 165f, 184, 185

crop prices: 1920s and 1930s, 112; 1950s–present, 115–116, 126, 188

crop residues: after conventional tillage, 134–135; for soil health, 167, 168–169

crop rotation: of corn and soybeans, 117; in regenerative agriculture, 169, 171, 190, 191, 196

crop yields: and climate change, 187, 189; financial losses, 133; government farm policy, 192; from loss of topsoil, 132. *See also* corn; soybeans

cropland: acres in Iowa, 126; artificial drainage of, 81; under conventional tillage, 134; in cover crops, 164; erosion rates, 5, 121, 122f, 125–128, 130; under no-till, 167; prevented planting of, 188; sale price and CSR2, 53

crops: historically planted in Iowa, 109, 110. *See also* corn; cover crops; soybeans

CSR2, 52–53. *See also* Corn Suitability Rating

cut-and-fill cycles. *See* Holocene alluvial cycles

Darwin, Charles, 38, 158

decomposition. *See* enzymes; microbes; organic matter

Deere, John, 108

derecho (2020), 196

Des Moines flash flood (2018), 182

Des Moines Lobe glacier: erosion east and west of, 95–96; multiple advances, 66; path from the north, 67f; stagnation, 80

Des Moines Lobe landform region: map, 65f, 67f; paleoenvironments, 68–69; prairie potholes, 80–81; soil parent materials, 66, 78; wetlands and carbon sequestration, 194

Dietzel, Kevin, 178

Dietzel, Ranae, 178

Dig It! Secrets of the Soil (exhibit), 142

dirt, 28, 161

Dirt, the Erosion of Civilizations (Montgomery, 2008), 120

Dokuchaev, Vasily: *Russian Chernozem*, 38

"Don't Treat It Like Dirt" (YouTube), 28

Downs soil series, 82; profile, 83f

drainage districts: creation of, 111

drainage tile. *See* artificial drainage

drought: prehistoric on Des Moines Lobe, 69, 97; twentieth century, 112–116; since 2000, 126, 190, 191; and grazing, 177; projected, 188

Dust Bowl, 112

dynamic soil properties: defined, 27, 63. *See also* bulk density; organic carbon; soil color

E horizon, 29f, 30, 45, 46f, 82, 83f

earthworm casts. *See* earthworms

earthworms, 74, 101f, 149, 158–159, 169; casts, 158; casts in paleosols, 100, 101f; and gardening, 184; organic matter shredders, 158–159

Entisols, 42, 43f; from erosion, 70–71, 100, 106; profile, 46, 47f, 71f

environmental quality. *See* climate change effects on farming; human impacts on soil; runoff; soil health; water quality

Environmental Working Group, 127. *See also* EWG Midwest

enzymes: role in decomposition, 155

ephemeral gullies, 128–134, 129f; defined, 128

erosion: following settlement, 1–2, 5, 47, 48, 70–71, 106; historic rates, 113–114, 119–120; modern rates, 121–128, 122f; and slope position, 23f. *See also* channelization, historic; ephemeral gully; sheet and rill erosion

erosion prevention. *See* cover crops; crop residues; no-till

ethanol from corn, 119, 126

Euro-American settlement: erosion caused by, 70; history of, 107–110

Eutrudepts, 49

EWG Midwest, 127–130, 132

expandable clay, 76, 90

exudates. *See* roots

family farms, 172, 179; decline of, 5, 115, 196

farm bill. *See* government farm policy

farm chemicals: in twentieth century, 114, 140, 141; in waterways, 137–138. *See also* herbicides; inorganic nitrogen; pesticides

Farm Crisis (1980s), 115–116

farm equipment, 113–117, 134–135, 165–166, 165f, 173; as cause of soil compaction, 135–136, 183

farms: decline in numbers, 115–116; increase in size, 116, 196; number of organic, 141. *See also* family farms

fauna as soil-forming factor, 68. *See also* arthropods; earthworms

Fayette soil series, 45–46, 69, 74, 77; profile, 45f

Fenton, Thomas, 52

fertilizers: current use of, 6, 114, 117, 140, 155–156; historic use of, 5, 114; and water pollution, 181. *See also* inorganic nitrogen

food security: and population growth, 143, 192; role of healthy soils, 197

Food Security Act of 1985, 125, 126

forests: history in the Midwest, 69.

See also vegetation as soil-forming factor

fossil record: Quaternary, 68

Fukuoka, Masanobu, 141

fungi, 21, 150, 152–154, 153f, 163. *See also* mycorrhizal fungi

Future Farmers of America, 12

gardens/gardening: rates of pesticide use, 181; and soil health, 142, 152, 159, 161, 183–184

genetically modified seed, 116–117

geologic dating. *See* cosmogenic dating; fossil record; magnetic polarity of Earth; radiocarbon dating; Yellowstone ash layers

geologic erosion: Holocene, 69, 70, 100; Iowan Erosion Surface, 66, 95; paleosols, 83, 86–87, 94–95, 97

glacial deposits: map, 65f; thickness in Iowa, 91

glacial erratics, 95

glacial kettles, 80, 81

glacial till: bulk density, 27; C horizons in, 35, 94; composition, 66, 75; units in Iowa, 78, 91–92, 93f

glaciation in Iowa. *See* Illinoian glaciation; pre-Illinoian glaciation; Wisconsinan glaciation

gleyed soil: defined, 78

Glinka, Konstantin, 38

global food supply, 115. *See also* food security

global warming. *See* climate change effects on farming; greenhouse gases

glomalin, 154

glyphosate, 185, 189

Gosport soil series, 24

government farm policy: Agricultural Extension Act of 1906, 111; conservation compliance, 125, 130–131; Conservation Reserve Program, 125, 126, 195; crop insurance, 164, 173; farm bill, 125–126, 164, 167; Food Security Act of 1985, 125, 126; history of, 112; leading to overproduction, 192; prevented planting, 188; Soil Conservation Act (1935), 120

granular structure, 20f, 21f, 74; formation of, 20, 157, 179; in paleosols, 96, 101f

grassed waterways, 131f, 132; defined, 131

grazing: and drought, 190–191; effect on soil, 176–177, 178; history in Iowa, 109, 111; loss of organic matter under continuous grazing, 177; manure as organic matter, 171; prescribed grazing, 177–178, 191

Great Depression, 112–113

Green Revolution, 115, 119, 140, 143

greenhouse gases, 191, 194. *See also* carbon dioxide; methane

groundwater: behavior of, 77; seep, 87, 88f. *See also* soil drainage class

Growing a Revolution (Montgomery, 2017), 142

gullies. *See* classic gullies; ephemeral gullies

gumbo, 87

Haplosaprist, 50

herbicides, 165, 167, 185; pollution from use on lawns, 181. *See also* glyphosate

highly erodible land: defined, 125; and ephemeral gully erosion, 132
Highway Guide of Iowa Soil Associations, 68
Hilgard, E. W., 38
Histosols, 42, 43f, 49–51, 80; profile, 49f
Hole, Francis D., 7, 14, 161
Holocene alluvial cycles, 71, 99–100
Holocene Epoch: paleoenvironments, 69
Holocene paleosols, 86–87, 98–100
horizons. *See* soil horizons
Houghton soil series, 49–51; profile, 49f
Howard, Albert, 141
human impacts on soil, 105–106; summary of, 117–118. *See also* compaction; erosion; organic carbon
humus, 5, 148, 194; defined, 147
hydric soils, 81
hyphae, 152, 153f

ice-age mammals, 68
ice ages in Iowa. *See* Illinoian glaciation; pre-Illinoian glaciation; Wisconsinan glaciation
Ida soil series, 70; profile, 71f
Illinoian glaciation, 93, 93f, 94, 97
Inceptisols, 42, 43f; profile, 48f, 48–49, 70
Indian corn, 109
infiltration: under conventional tillage, 135, 137f, 138; with cover crops, 164, 170, 184; with prescribed grazing, 177–178; prior to settlement, 148; and soil development,

73; in urban soils, 182–183; YouTube demonstration, 183. *See also* runoff
Ingham, Elaine, 144
inorganic nitrogen, 76, 114, 117, 138, 156
insects: beneficial, 170
interglacial times: defined, 86; landscape processes and soil formation, 94, 96
invertebrates, 159. *See also* earthworms
Iowa Agricultural Experiment Station, 39, 112, 120
Iowa counties: Adair, 95, 113; Adams, 32, 33f, 88, 89f, 100; Allamakee, 39, 99f; Appanoose, 13f, 46f, 89; Audubon, 56, 57; Black Hawk, 56, 57; Boone, 79f; Bremer, 39, 159; Cass, 95; Cerro Gordo, 39, 49; Clarke, 89, 100; Davis, 170f, 191; Decatur, 53, 86; Dubuque, 56; Fayette, 131f; Guthrie, 133; Hamilton, 44, 81, 178, 180f; Hancock, 49; Henry, 171; Ida, 110; Johnson, 182; Keokuk, 86; Kossuth, 111f; Lucas, 89; Lyon, 64; Madison, 31, 100, 176; Mahaska, 20, 21f; Marion, 133; Marshall, 59f, 97f; Monona, 51, 71f; Polk, 182; Pottawattamie, 47f, 95; Poweshiek, 44; O'Brien, 53; Sac, 56, 110; Shelby, 172, 174, 174f; Sioux, 109f; Story, 39, 44, 48f; Tama, 14, 39, 108; Union, 86, 91, 177; Washington, 168; Wayne, 89; Winnebago, 49; Winneshiek, 45f, 83; Worth, 49, 54
Iowa Daily Erosion Project, 127, 128
Iowa Division of Soil Conservation, 55, 121, 132

Iowa Farmers Union, 175, 196
Iowa Interfaith Power and Light, 192
Iowa Learning Farms, 196
Iowa state soil, 11
Iowa State University Extension and
    Outreach: erosion calculator, 124;
    Iowa Farm and Rural life Poll, 164;
    soil mapping, 55
Iowan Erosion Surface, 66, 95; map,
    65f
Iowa's Geological Past (Anderson,
    1998), 64
iron oxides, 26f. *See also* oxides and
    oxidation

Jenny, Hans, 37

Krotovina: in paleosols, 100
Kyoto Protocol (1997), 191. *See also*
    carbon sequestration by soils

Lake Panorama, 133
Lake Red Rock, 133
Lamoni soil series, 52–53, 89–90
land stewardship, 197
land use, 2; acres of cropland, 126;
    acres under cover crops, 164; acres
    of developed land, 180; acres of
    pastureland, 176; acres under no-
    till, 167; impact of soil texture, 16
land value: in 1800s, 107–108; in
    twentieth century, 111; collapse of
    in 1970s–1980s, 115–116; and rental
    rates, 53; and soil series, 52
landform regions of Iowa, 65f
Landforms of Iowa (Prior, 1991), 65
Laurentide ice sheet, 65, 92

Law of Return, 141
leaching, 33f; 75–76; defined, 33
Leopold Center for Sustainable Agri-
    culture, 196
LiDAR, 58–59
*Life in the Soil: A Guide for Naturalists
    and Gardeners* (Nardi, 2007), 159
lime (soil amendment): soil pH, 76
linear extensibility rating: defined,
    90; and paleosols, 90
livestock. *See* grazing
loam: defined, 17; classes of, 18; in
    thin section, 19f
Lobe, James, 66
loess: defined, 3–4; bulk density, 27;
    Corn Suitability Rating, 52–53;
    linear extensibility rating, 90; loess
    kindchen, 34; and paleosols, 94,
    97; as parent material, 11, 12f, 15,
    24, 26f, 75; source of, 3, 19; texture,
    18, 19f, 27; thickness, 65f, 66–68.
    See also Loveland Loess; Peoria
    Formation
loess erosion, 30, 47, 70, 71f, 72f, 100,
    106, 123. *See also* tolerable soil loss
Loess Hills: erosion, 47, 70, 71f, 123;
    map, 65f, 66
Losing Ground (EWG Midwest 2011
    report), 127–128
Lost Lake Farm, 178–180, 180f
Loveland Loess, 93f, 94
Lowdermilk, W. C., 1, 119–120
Lyle, Levi, 164, 165f, 166

macronutrients. *See* plant nutrients
macropores: defined, 21; and soil
    aeration, 135; in thin section, 75f

magnetic polarity of Earth, 92
Major Land Resource Areas, 54
mammals: ice age, 68; and soil forma-
    tion, 159
manganese. *See* minerals in soil
manganese oxides. *See* oxides and
    oxidation
manure: in gardening, 184; in regen-
    erative agriculture and organic
    farming, 169, 171, 172, 173; in
    prescribed grazing, 179; and water
    quality problems, 179
mapping Iowa soils. S*ee* soil maps
    and mapping
Marbut, Curtis, 38–39
methane, 180, 191, 193
microbes, 74, 142, 149, 184, 190. *See
    also* fungi; nematodes; protozoa;
    soil organisms
micronutrients. *See* plant nutrients
micropores: defined, 21
Middle Raccoon River watershed,
    133
Miller, Bradley, 58–59
Miller, Gerald, 40
Miller, Thom, 171
minerals in soil, 19f, 32f, 75–77; effect
    on soil color, 23; as plant nutrients,
    145–146, 155; weathering of, 76. *See
    also* expandable clays; iron oxides;
    oxides and oxidation
Minger, Mark, 56–57
Mississippi River, 93, 137
Missouri River, 137
moldboard plow, 108, 109f, 114, 117, 134
Mollisols, 3–4, 4f, 42–44, 43f; A hori-
    zon, 3, 44–45, 78, 88; formation of,

73–74, 75; and prairie restoration,
    194, 195; profile, 12f, 44, 79f
Monsanto. *See* Bayer (company)
Montgomery, David, 120
Munsell Soil Color charts, 24, 25f
mycorrhizal fungi, 153–154, 157, 163,
    192

National Climate Assessment, 187
National Cooperative Soil Survey, 39,
    41, 58
National Resources Inventory, 121,
    125, 127, 129, 130
National Soil Survey Center, 54
Native Americans, 86–87; and agri-
    culture, 107
Natural Resources Conservation
    Service, 120–121; erosion estimates,
    123, 124, 128; grassed waterways,
    132; Major Land Resource Areas,
    54–55; National Resources Inven-
    tory, 121; soil health initiative, 144;
    soil mapping in Iowa, 55
nematodes, 156, 157f
Nicollet soil series, 78–79; profile, 79f;
    and water table, 79–80
night crawlers. *See* earthworms
nitrate/nitrogen: nitrous oxide emis-
    sions, 193; as plant nutrient, 107,
    114, 145, 146, 151, 155, 156, 190; water
    pollution, 6, 118, 137–138, 149, 156,
    189, 190
Nodaway soil series, 47, 70; profile,
    47f
nodules: 32, 33f; defined, 77
no-till, 137f; acres farmed in Iowa,
    167–168; and earthworms, 159; and

improved infiltration, 190; weed reduction, 173

O horizon, 29f, 30, 49f, 50
Okoboji soil series, 80
organic carbon: carbon sequestration, 191–195; dynamic soil property, 27, 63; loss due to agriculture, 193, 195; in Mollisols, 44; percent of organic matter, 30
organic farms and farming, 141, 172–173
organic matter: decomposition by organisms, 146, 155; historic loss of, 5; percent in topsoil, 25, 30; role in ped stability, 148; and soil color, 23–25; source in prairies and forests, 30, 74, 149; three main types, 147–148; water infiltration and absorption, 138, 148. *See also* cover crops; earthworms; Law of Return; soil organisms
organic nitrogen, 156
organic soils, 42
organic-ish farming, 142, 179
oxides and oxidation, 23–25, 26f, 31–34, 76–77; in coatings, 75f, 76

paha, 95
paleosols: in bedrock, 86; clues to paleoenvironments, 95–98; groundwater behavior, 88f, 88–89, 90; Holocene, 86–87, 98–100; parent material for modern soils, 89–90; Pleistocene, 86, 89, 91–93, 94–98; profiles, 89f, 97f, 99f
palimpsest soil, 82, 88, 98

parent material as soil-forming factor, 64–68; and bulk density, 27; effect on soil color, 24; and soil texture, 16, 19
Paris Climate Agreement (2015), 191
pedogenesis. *See* soil formation
pedogenic carbonates, 32–34, 33f, 49; depth of, 80, 96. *See also* leaching
pedology. *See* soil classification; soil formation
peds, 19, 21, 51f, 101f; and carbon sequestration, 195; damaged by tillage, 135; and organic matter, 147, 148; stability of, 27, 63, 154, 158, 162, 164. *See also* soil structure
Peoria Formation, 65–66. *See also* loess
permafrost, 66
pesticides, 6, 114, 117, 140, 154; on lawns and gardens, 181
pH (of soils), 76; defined, 63
phosphorus: plant nutrient, 145; pollution, 6, 118, 137, 149
photosynthesizers, 150, 162
plant and animal fossils, 68
plant nutrients, 145–146, 155
Pleistocene stratigraphy in Iowa, 91, 92, 93f, 94–95
plow pan, 136
plowshare: defined, 108
pollinators: benefits of cover crops, 170, 175
pollution, water. *See* herbicides; nitrate/nitrogen; phosphorus; runoff; sedimentation; water quality
population growth, 110, 115, 192, 197

porosity. *See* soil porosity
postsettlement alluvium, 99f, 100
potatoes: history in Iowa, 109
Practical Farmers of Iowa, 166, 173,
    174f, 175, 186, 196
prairie plants: organic matter, 74, 149;
    root penetration, 73, 149
prairie potholes: in north-central
    Iowa, 43, 50–51, 50f, 80–81; wetland
    restoration, 194
prairie soils, 3, 39, 42–44, 73, 74, 78.
    See also Mollisols
prairie strips, 149
prairies, 3–4, 30, 69; acreage remain-
    ing, 105; and carbon sequestration,
    195; conversion to cropland, 70, 82,
    106, 108–110, 109f; restoration of,
    194, 195; water retention in, 148
precipitation, extreme: effects on
    water quality, 189–190; predictions
    for, 190
Pre-Illinoian glaciation, 91–95, 97,
    101f; till layers, 64–65, 93, 93f, 94,
    95. *See also* paleosols
prescribed grazing, 176–180
profile. *See* soil profile
protozoa, 154–156

Quaternary Landscapes in Iowa
    (Ruhe, 1969), 23

Rabinowitz, Ruth, 175
radiocarbon dating, 92, 98, 100
railroad construction in nineteenth
    century: effect on agriculture, 110
rainfall: and erosion rates, 127–129;
    impact on soil peds, 135. *See also*
    infiltration; precipitation, extreme

Rainscaping Iowa, 182
redox depletions, 77–78
redox features: defined, 77. *See also*
    oxides and oxidation
reduction (chemical), 24, 46, 77
regenerative agriculture, 161, 163–176,
    187, 192–197; and gardening,
    183–184; government policy, 197
rented cropland, 5, 174–175; rental
    rates and CSR2, 52–53. *See also* ten-
    ant farming
resilient farming, 195–197; for eco-
    nomic survival, 188; and weather
    unpredictability, 187, 196
Revised Universal Soil Loss Equation,
    124
rhizosphere, 151, 154
Riecken, Frank, 40, 41
rill erosion, 124f; defined, 124. *See also*
    sheet and rill erosion
riparian buffers, 121
Rodale, Jerome, 141
roller crimper, 165, 165f
roots: depth of, 70, 149, 183f; develop-
    ment of, 31; erosion prevention
    by, 164, 168; exudates, 147, 164, 171;
    of native and non-native grasses,
    183f; penetration of soil by, 63, 73,
    135–136, 149, 170; source of organic
    matter, 73–74
Rosmann, Maria, 172
Rosmann, Ron, 172
Roundup. *See* glyphosate
Ruhe, Robert: hillslope model, 23, 23f;
    Quaternary Landscapes in Iowa, 95
runoff: causes, 6, 22, 101, 135, 137, 137f,
    177; reducing, 138, 149, 162–164, 168,
    177; and sediment, 134; from urban

soils, 181; and water quality, 6, 118, 137–138, 148–149, 156, 181. *See also* infiltration

Russell, Matt, 192

Russian Chernozem (Dokuchaev, 1883), 38

sand: in soil texture, 14–18, 17f, 91. *See also* soil texture

Sangamon interglacial stage, 94

Sangamon paleosol, 93f, 97f, 97–98

Sass, Neil, 159

sedimentation, 123, 129, 129f, 134; dredging, 133

Seed Corn Gospel Train, 111

series. *See* soil series

Seventh Approximation (1960), 40

Sharpsburg soil series, 41, 52–53, 177–178

sheet erosion: defined, 124. *See also* sheet and rill erosion

sheet and rill erosion: caused by soil compaction, 137; daily calculations of, 127–128, 130; effect on water quality, 137; Iowa ranking in U.S., 126–127; rates, 121–128, 122f, 130

Shelby-Sharpsburg-Macksburg soil association, 41

shrink-swell potential. *See* linear extensibility rating

silt: from glacial grinding, 19; in soil texture, 14–18, 17f

Simonson, Roy, 40

slickensides, 33f, 34; in paleosols, 96

slope positions, 23f. *See also* topography as soil-forming factor

smectite. *See* expandable clay

Smith, Guy, 40

sodbusters, 108

soil association, 67–68; defined, 41

soil biology, 139, 146, 147–148, 155; roots, 74, 164, 171. *See also* arthropods; bacteria; earthworms; fungi; microbes

soil carbon: release by conventional tillage, 134. *See also* carbon sequestration by soils; organic matter; soil fertility

soil classification: history of, 37–42; by soil order, 41–52, 43f; by texture, 16–18, 17f. *See also* soil maps and mapping; *Soil Taxonomy*

soil color, 22–26; effect of groundwater, 77–78; effect of topography, 72; iron oxides, 76; low-chroma colors, 78; Munsell charts, 25f; in thin sections, 18

Soil Conservation Service. *See* Natural Resources Conservation Service

soil consistence: defined, 27

soil crust, 135, 137, 190

Soil Data Join Recorrelation, 54

soil degradation: summary, 117–118. *See also* compaction; tillage

soil density. *See* bulk density

soil drainage class: effect of bulk density, 26; effect on soil color, 24; on Des Moines Lobe, 80; and topography, 23f

soil ecosystem: defined, 149; sustainability and nutrient cycling, 155

Soil Erosion in Iowa (Walker and Brown, 1936), 120

soil fertility, 5, 27, 139–141, 145–147; loss of from tillage, 134–135; strategies for improving, 162, 163. *See also* topsoil

soil food web, 149–159, 150f

soil formation: defined, 17; processes, 28–35, 73–83. *See also* fauna as soil-forming factor; parent material as soil-forming factor; time as soil-forming factor; topography as soil-forming factor; vegetation as soil-forming factor

soil great groups, 41, 42, 44

soil health: defined, 142; grazing, 176; organic matter, 145–148, 155, 162; porosity, 148; principles of regenerative agriculture, 163; root systems, 149; soil organisms, 139, 151–154, 156–158; of urban soils, 180

soil horizons, 29, 29f, 31, 34–35, 73, 81; defined, 28. *See also* A horizon; B horizon; C horizon; E horizon; O horizon

soil loss. *See* ephemeral gullies; sheet and rill erosion; topsoil

soil maps and mapping, 53–60, 59f; history in Iowa, 39–41

soil memory, 63, 81

soil orders: defined, 41; distribution of, 43f; main orders in Iowa, 42; sources of taxonomic names, 42–43. *See also* Alfisols; Entisols; Histosols; Inceptisols; Mollisols; Vertisols

soil organisms: decline in, 140; numbers of, 149; role in carbon sequestration, 194; in soil food web, 150, 150f; and soil health, 139; soil ratings for aerobic organisms, 54

soil phases, 58, 59f, 72–73

soil porosity: defined, 16; damaged by tillage, 101, 135, 136f; as dynamic soil property, 27, 148; improving, 148–149, 158, 164, 171, 180f, 182

soil profile, 11–13, 29f; defined, 11

soil properties. *See* bulk density; soil color; soil consistence; soil fertility; soil porosity; soil structure

soil replacement rate: estimation of, 126. *See also* tolerable soil loss

soil series, 67–68; and cropland value, 52–53; defined, 13, 43; naming of, 13–14

soil structure, 19–22, 20f; A horizon, 21f; B horizon, 22f, 31, 44, 76; damage by conventional tillage, 135–138; E horizon, 45, 96; in paleosols, 96, 98f, 101f; and porosity, 148–149, 154. *See also* blocky structure; granular structure

soil suborders: defined, 42–43; Aqualf, 46; Aquoll, 42–43, 88; Udalf, 46; Udoll, 43, 44

soil surveys: defined, 13; contents of, 53–54; history of in Iowa, 39; publication of, 57. *See also* soil maps and mapping

Soil Taxonomy (1975, 1999), 41, 55

soil texture, 14–19, 17f, 27

Soils of Iowa (Brown, 1936), 39

Southern Iowa Drift Plain, 65f, 89

soybeans, 5, 165f; and cover crops, 165, 168, 185; introduction of, 112;

in no-till, 167; organic, 173; prices, 126; yields, 116, 119, 171

Spirit Lake, 80

Steele, Jason, 171–172

stepped landscape model, 95

stone line, 95

Storden soil series, 48–49; profile, 48f

Storm Lake, 80

stormwater: management of, 182

strip cropping, 121

structure. *See* soil structure

subsoil: colors of, 25, 26; defined, 20; structure of, 22f

suicide among farmers, 116, 188

sustainable agriculture. *See* regenerative agriculture

Symphony of the Soil (film), 142

synthetic farm chemicals. *See* farm chemicals

T value. *See* tolerable soil loss

Tama soil series, 11, 30, 43–44; bulk density, 27; colors, 14, 25; composition, 15; drainage, 77; profile, 12f; topsoil, 20

Teaming with Microbes (Lowenfels and Lewis, 2010), 142

temperature: crop yields, 189; soil microorganisms, 190; weed growth, 189. *See also* climate change effects on farming

tenant farming, 5–6, 133, 173–176

terraces (conservation practice), 121, 132

texture. *See* soil texture

thin sections: for study of soil, 18, 82; carbonate nodules, 33f; clay coatings, 32f, 75f; porosity, 136f, 180f; soil structure, 21f, 22f, 97f, 101f, 180f; soil texture, 19f

Thompson, Dean, 100

tillage: and soil degradation, 134–138

tilth, 166; defined, 4; declining, 117

time as soil-forming factor, 62–63, 73–83

tolerable soil loss, 125

topographic maps, 59

topography as soil-forming factor, 70–73; effect on soil color, 23

topsoil, 4f; abundance of microbes in, 150, 151, 154; building topsoil with cover crops, 171; color, 25; degradation of, 27, 101, 136f, 166, 181; erosion of, 58, 71, 119, 123, 124f, 126, 129f; original thickness in Iowa, 3, 5; soil-forming processes, 73, 76; soil structure, 20, 21f, 179, 180f. *See also* A horizon; compaction; ephemeral gullies; organic matter; sheet and rill erosion

transformation: in soil formation, 76

transitional horizon, 34–35, 45

translocation: in soil formation, 74, 75, 76

tundra: defined, 68

Udifluvent, 47

Ultisols, 62

Understanding Iowa Soils (Simonson et al., 1952), 40

Universal Soil Loss Equation, 124

University of Iowa Museum of Natural History, 68

updating of soil maps, 68. *See also*
    soil maps and mapping
Upper Midwest: history of vegetation,
    68–69, 70; topography of, 67f
urban soils, 181–186
U.S. Department of Agriculture:
    Farm Service Agency aerial pho-
    tography, 56; soil classification, 16;
    soil health initiative, 143–145; soil
    mapping, 55. *See also* government
    farm policy; National Cooperative
    Soil Survey; Natural Resources
    Conservation Service

vegetation as soil-forming factor,
    68–69. *See also* prairies; prairie
    soils
vertebrates: in soil food web, 159
Vertisols, 42, 43f; profile, 51f, 51–52
Vilsack, Tom, 144

Washout (EWG Midwest 2013 report),
    130
water quality: and cover crops, 164;
    and erosion, 123, 137; in Iowa, 123,
    148, 156, 179, 190, 194; and urban
    soils, 181
water saturation: and redox features,
    77–78, 80. *See also* soil drainage
    class
water table: and artificial drainage,
    81; depth of: 77–80, 123; on Des
    Moines Lobe, 80; and soil color, 24
weather: defined, 188; unpredictabil-
    ity, 187–189. *See also* climate change

effects on farming; drought; pre-
    cipitation, extreme
Web Soil Survey and WebSoilSurvey.
    gov, 57, 58, 90
Webster soil series, 80
weed control: with conventional till-
    age, 172–173; with no-till, 167
wetlands: carbon emissions from
    drainage of, 193; carbon sequestra-
    tion potential, 194; conversion to
    farmland, 110, 111; restoration, 194
wheat: history in Iowa, 109
wind erosion, 123; 1930s, 112; 1950s, 114
windbreaks, 121
Wisconsinan glaciation, 93–95, 93f,
    97; multiple advances of, 64
Wisner, Robin, 55–56, 57
Wojda Century Farm, 18, 163
Women, Food and Agriculture Net-
    work, 175, 196
woodland: conversion to pasture-
    land, 70; conversion to farmland,
    110
woodland soils, 42
World Reference Base for Soil Re-
    sources, 105
World Soil Day, 142
World War I, 112
World War II, 39, 113

Yarmouth interglacial stage, 94
Yarmouth-Sangamon paleosol, 93f, 94
Yellowstone ash layers, 57, 92
young soils, 31, 34, 63. *See also* Enti-
    sols; Inceptisols

Other Bur Oak Books of Interest

A Bountiful Harvest:
The Midwestern Farm Photographs
of Pete Wettach, 1925–1965
by Leslie A. Loveless

A Country So Full of Game:
The Story of Wildlife in Iowa
by James J. Dinsmore

Ecological Restoration in the Midwest:
Past, Present, and Future
edited by Christian Lenhart
and Peter C. Smiley Jr.

The Ecology and Management of
Prairies in the Central United States
by Chris Helzer

The Emerald Horizon:
The History of Nature in Iowa
by Cornelia F. Mutel

Fragile Giants:
A Natural History of the Loess Hills
by Cornelia F. Mutel

Gardening in Iowa and
Surrounding Areas
by Veronica Lorson Fowler

Gardening the Amana Way
by Lawrence L. Rettig

Green, Fair, and Prosperous:
Paths to a Sustainable Iowa
by Charles E. Connerly

An Illustrated Guide to
Iowa Prairie Plants
by Paul Christiansen
and Mark Müller

Iowa's Archaeological Past
by Lynn M. Alex

Iowa's Geological Past:
Three Billion Years of Earth History
by Wayne I. Anderson

Iowa's Minerals:
Their Occurrence, Origins,
Industries, and Lore
by Paul Garvin

Land of the Fragile Giants:
Landscapes, Environments,
and Peoples of the Loess Hills
edited by Cornelia F. Mutel
and Mary Swander

Landforms of Iowa
by Jean C. Prior

My Vegetable Love:
A Journal of a Growing Season
by Carl H. Klaus

Of Men and Marshes
by Paul L. Errington

Of Wilderness and Wolves
by Paul L. Errington

Out Home
by John Madson
and Michael McIntosh

A Practical Guide to Prairie
Reconstruction
by Carl Kurtz

Restoring the Tallgrass Prairie
by Shirley Shirley

Stories from under the Sky
by John Madson

A Sugar Creek Chronicle:
Observing Climate Change from
a Midwestern Woodland
by Cornelia F. Mutel

The Tallgrass Prairie Reader
edited by John T. Price

Twelve Millennia:
Archaeology of the Upper
Mississippi River Valley
by James L. Theler
and Robert F. Boszhardt

Up on the River:
People and Wildlife of the
Upper Mississippi
by John Madson

The Vascular Plants of Iowa:
An Annotated Checklist and
Natural History.
by Lawrence J. Eilers
and Dean M. Roosa

A Watershed Year:
Anatomy of the Iowa Floods of 2008
edited by Cornelia F. Mutel

Where the Sky Began:
Land of the Tallgrass Prairie
by John Madson